Emerging Equine Science

KW-000-797

BISHOP BURTON LRC
WITHDRAWN

ACCESSION No. TO46713

CLASS No. 636.1089 Ref

Emerging Equine Science

Edited by

J. Alliston, S. Chadd, A. Ede, A. Hemmings, J. Hyslop,
A. Longland, H. Moreton and M. Moore-Colyer

NOTTINGHAM
University Press

Nottingham University Press
Manor Farm, Main Street, Thrumpton
Nottingham, NG11 0AX, United Kingdom

NOTTINGHAM

First published 2004
© British Society of Animal Science

All rights reserved. No part of this publication
may be reproduced in any material form
(including photocopying or storing in any
medium by electronic means and whether or not
transiently or incidentally to some other use of
this publication) without the written permission
of the copyright holder except in accordance with
the provisions of the Copyright, Designs and
Patents Act 1988. Applications for the copyright
holder's written permission to reproduce any part
of this publication should be addressed to the publishers.

British Library Cataloguing in Publication Data
Emerging Equine Science
I. Allison, J., II. Chadd, S., III. Ede, A., IV. Hemmings, A., V. Hyslop, J., VI. Longland, A., VII.
Moreton, H., VIII. Moore-Colyer, M.

ISBN 1-897676-47-6

Disclaimer

Every reasonable effort has been made to ensure that the material in this book is true, correct,
complete and appropriate at the time of writing. Nevertheless, the publishers and authors do
not accept responsibility for any omission or error, or for any injury, damage, loss or financial
consequences arising from the use of the book.

Typeset by Nottingham University Press, Nottingham
Printed and bound by Hobbs the Printers, Hampshire, England

CONTENTS

B Essén-Gustavsson
*Department of Large Animal Clinical Sciences, Unit for
Comparative Physiology and Medicine, Faculty of Veterinary
Medicine, Swedish University of Agricultural Sciences,
Box 7018, 75007 Uppsala, Sweden*

J L L Rivero
*Laboratory of Muscular Biopathology, Department of
Comparative Anatomy and Pathological Anatomy, Faculty
of Veterinary Science, University of Cordoba, Cordoba, Spain*

J F Burn
*Department of Anatomy, University of Bristol, Southwell Street,
Bristol, BS2 8EJ, UK*

D Marlin
*Head of Physiology, Animal Health Trust, Newmarket,
Suffolk and Visiting Professor of Cardiovascular and
Respiratory Physiology, University of Bristol*

S D McBride[1] and A Hemmings[2]
*[1] Institute of Rural Studies, University of Wales, Aberystwyth,
SY23 3AL, UK; [2] Royal Agricultural College,
Cirencester, Gloucestershire, GL7 6JS, UK*

HRH the Princess Royal speaking at the Emerging Equine Science conference

BUCKINGHAM PALACE

I am very pleased to be able to introduce 'Emerging Equine Science' based on the very first British Society of Animal Science (BSAS) conference on horses held at the Royal Agricultural College, Cirencester in September 2003. BSAS, through this book, is expanding its remit away from farm animals and, by including horses, recognises that this animal is a very important and expanding part of UK rural society.

It is quite common for those that look after animals to continue to do things in a way that they and their predecessors have always done. This isn't always a sound policy and this book is a first real effort to bring together ideas and views from many scientists, advisors and keepers of horses on the science that lies behind improving the keeping and care of horses.

The UK Equine Industry continues to flourish as an important part of the rural community. However, although significant advances have been made in the science that underpins the industry, there is a lack of research funds for equine science. This is frustrating as the industry doesn't neatly fall into a 'research category' and is shunned by Agricultural or Social Research Funders. Yet an enhanced understanding of the science should mean we are able to treat better the horses used in businesses such as racing, eventing and those associated with horse-riding for pleasure.

Because many horses are kept as companions to humans in Britain there is a tendency to treat them rather like children. This is a mistake and there is much to be done in understanding how our care of horses should change to meet better their needs in terms of nutrition, health and behaviour.

The problems of poor health and inappropriate nutrition in equines stem from two aspects; a lack of understanding of the basic biology that underpins health, and a focus for too long on treating chronic disease rather than proactive health care. Changing the emphasis to gaining a better understanding of health promotion will provide a sound basis for equine management that can help cast aside practices based on myth.

We currently do not understand how to keep horses in confined spaces. Horses, if conditions are reasonable, are often more comfortable outside in the company of other horses as they are naturally herd animals. It is now recognised that one of the challenges the industry faces is to understand the conditions that will help horses live in harmony in the ones and twos that is current practice in much of this country.

Behavioural problems are in fact one of the very visible symptoms of this lack of harmony and innovative research from the UK is showing that very different approaches may be the key to success with horses demonstrating stereotypical behaviour. For example it is now evident that prevention of 'contagious vices' such as crib-biting may adversely affect a horse's welfare and its ability to cope with stressful situations.

'Emerging Equine Science' is the result of the co-operation of a wide cross section of those involved in research, education, societies, advisory organisations and the Equine Industry. The need to get good scientific information to those involved at all stages of the industry is well recognised, but current networks are fragmented. New approaches and interactive networks that allow easy access to information based on sound science are essential. I hope that you find this first book on 'Emerging Equine Science' a useful source of information and a stimulus to inquiring of others how better we can all look after horses.

H R H The Princess Royal

1

Metabolic responses of muscle to exercise

B Essén-Gustavsson
Department of Large Animal Clinical Sciences, Unit for Comparative Physiology and Medicine, Faculty of Veterinary Medicine, Swedish University of Agricultural Sciences, Box 7018, 75007 Uppsala, Sweden

Abstract

Muscle is a tissue with a great plasticity due to the fact that it is composed of fibres having different contractile and metabolic properties. In horses, muscle metabolic responses to exercise are studied by taking biopsies from the gluteus medius muscle. Histochemical stains are used to identify slow contracting type I fibres and fast contracting type IIA and type IIB fibres and to evaluate fibre areas, capillary supply, oxidative capacity, glycogen and lipid content in a muscle. Biochemical analyses of substrates, metabolites and enzyme activities are performed either on a whole piece of muscle, on pools of fibres or on single fibres of identified type.

All fibres contain glycogen whereas lipid is mainly found in type I and type IIA fibres that have smaller cross-sectional areas and a higher oxidative capacity than type IIB fibres. Large variations can be seen in metabolic profile between and within fibre types. The most common muscular adaptation to training is an increase in oxidative capacity, capillary density and an increase in the type IIA/IIB ratio. The order of recruitment of fibres during most types of exercise is from type I to type IIA and type IIB.

The higher the intensity of exercise, the faster is the breakdown of glycogen. After racing (1640-2640m), and after high intense treadmill exercise, concentrations of lactate and inosine monophosphate (IMP) are increased in the muscle and concentrations of glycogen, adenosine triphosphate (ATP) and creatine phosphate (CP) decreased. Extremely low ATP and high IMP concentrations especially in some type II fibres are observed after racing.

After exercise of low intensity and long duration glycogen and triglyceride stores in muscle are utilised, amino acid metabolism is enhanced and protein degradation may occur. After submaximal treadmill exercise to fatigue and after endurance rides glycogen is degraded and depletion occurs mainly in type I and type IIA fibres.

Fibre type composition, substrate sources and differences in metabolic properties among fibres and the extent to which fibres are recruited are all factors that influence the metabolic responses of muscle to exercise. Biochemical analyses on whole muscle must be interpreted with caution since large variations in metabolic response to exercise occur among different fibres.

Introduction

Muscle contraction requires a continuous supply of ATP since the contractile mechanism results in hydrolysis of ATP and production of adenosine diphosphate (ADP) and free phosphate. The small stores of ATP and CP are the immediate sources for energy release in the muscle fibres. These stores can only provide energy for a short time and energy must therefore continuously be regenerated in the muscle from other sources. Carbohydrates and lipids, i.e. blood-borne glucose and free fatty acids and glycogen and triglycerides stored in the muscle are the major substrates for resynthesis of ATP.

Glycogen is the most important substrate for energy release, by anaerobic metabolism that often is associated with high lactate production. Anaerobic metabolism provides rapid but limited energy release. If oxygen is available energy is released by aerobic metabolism that is slower but yields much more ATP than anaerobic metabolism. ATP can also be produced from ADP and AMP is then deaminated to IMP and ammonia in the AMP deaminase reaction. This pathway is important for maintaining high ratios of ATP to ADP during periods when ATP utilisation exceeds supply.

Muscle aerobic and anaerobic capacity and metabolic responses to exercise have been investigated on the gluteus medius muscle because of its major role in locomotion and easy accessibility using the muscle biopsy technique (Lindholm and Piehl, 1974). This technique has made it possible to do both cross-sectional and longitudinal studies on horses. Muscle biopsies have mainly been obtained from Standardbred trotters, Thoroughbreds and horses participating in endurance rides. These studies show that muscle is a tissue with a great plasticity due to the fact that it is composed of fibres having different contractile and metabolic properties.

Muscle fibre properties and recruitment

Three main fibre types, slow contracting type I fibres and fast contracting type IIA and IIB fibres, are identified in horse muscle with histochemical staining for myosin ATPase activity (Essén *et al.*, 1980). Fibre sizes, capillary supply, oxidative capacity, glycogen and lipid content within the fibres are also evaluated with histochemical stains (Essén *et al.*, 1980; Essén-Gustavsson *et al.*, 1984; Karlström *et al.*, 1991). Histochemical stains are only semi-quantitative and a better estimate of oxidative capacity, glycogen and lipid content and muscle metabolic responses to exercise can be obtained from biochemical analyses. The biochemical analyses can either be performed on a whole piece of muscle or on pools of fibres or on single fibres that have been identified as type I, IIA or IIB with ATPase staining (Valberg and Essén-Gustavsson, 1987; Essén-Gustavsson *et al.*, 1997).

Type I and IIA fibres usually have smaller fibre sizes, higher oxidative capacity and lipid content than type IIB fibres but large variations can be seen between and within each fibre type (Essén-Gustavsson *et al.*, 1984; Valberg and Essén-Gustavsson, 1987; Valberg *et al.*, 1988). This is related to both hereditary factors as well as factors like age, sex and training status. Older and trained horses have a higher type IIA/IIB ratio, a higher oxidative capacity and a higher capillary supply as compared to younger and untrained horses (Essén *et al.*, 1980; Roneus *et al.*, 1992). Capillarisation of muscle fibres is of importance, both for delivery of oxygen and blood-borne substrates to each fibre and for the elimination of lactate and other metabolites. The most common muscular adaptation to training is an increase in oxidative capacity, capillary density and an increase in the type IIA/IIB ratio (Lindholm *et al.*, 1983; Essén-Gustavsson *et al.*, 1989; Roneus *et al.*, 1993).

An important factor for the metabolic response in muscle during exercise is the recruitment pattern of the muscle fibres. Histochemical staining for glycogen content in the fibres after varying intensities and durations of work indicates how different fibres are recruited (Valberg, 1986; Gottlieb, 1989; Essén-Gustavsson *et al.*, 1991). It is shown that the different fibres are recruited in the order from type I, to type IIA and type IIB. Thus, as the intensity of exercise increases more and more type IIB fibres get recruited. A consequence of this is that there will be a greater reliance on anaerobic metabolism with lactate production, as type IIB fibres usually show the lowest oxidative capacity among the fibres (Essén *et al.*, 1980; Valberg *et al.*, 1987; Valberg *et al.*, 1988). If horses perform incremental draught-loaded exercise during low and moderate intensity exercise both type I, IIA and type IIB fibres are recruited and ATP, CP and glycogen concentrations are decreased and lactate concentration increased after such exercise (Gottlieb *et al.*, 1988; Gottlieb, 1989).

Metabolic responses to high intensity exercise

During exercise glycogen is broken down in the fibres and the rate of breakdown is dependent on the intensity of the exercise. The higher the intensity of exercise the faster is the breakdown of glycogen and reliance on anaerobic metabolism and lactate production (Gottlieb, 1988; Valberg et al., 1989; Essén-Gustavsson et al., 1999).

Muscle biopsies have been obtained from horses after racing, before and after high intensity treadmill exercise to fatigue, which is when the horse cannot keep pace with the treadmill any longer and after near maximal treadmill exercise (Essén-Gustavsson and Valberg, 1987; Valberg et al., 1989; Essén-Gustavsson et al., 1995; Essén-Gustavsson et al., 1997; Schuback and Essén-Gustavsson, 1998). Results from these studies show that high intensity exercise leads to a rapid breakdown of glycogen with a concomitant production of lactate and a degradation of creatine phosphate. In addition, decreased ATP and increased IMP and ammonia concentrations are observed in the muscle after exercise and concentrations of ammonia, hypoxanthine and uric acid are increased in the blood. These metabolic changes show the importance of not only anaerobic glycolysis but also of the breakdown of CP and ATP and the AMP-deaminase reaction for fast energy production during short intense exercise. This also indicates a metabolic stress-situation since ATP production is insufficient to meet the demand.

Metabolic responses to high intensity exercise are shown to differ between horses and this is related to both fibre type composition and oxidative and glycolytic capacity of the muscle. It has been shown that horses with a high % of type IIB fibres and low oxidative capacity are more dependent on glycolysis and lactate production for their energy release during near maximal treadmill exercise (Valberg et al., 1985). In both racing thoroughbreds and standardbred trotters a high % of type IIB fibres is associated with a lower oxidative capacity and a higher glycolytic capacity and a greater lactate accumulation in muscle after racing (Valberg, 1987; Essén-Gustavsson et al., 1997). In one study the horses with the best racing times had the highest IIA/IIB ratio, the smallest type II fibre areas and the greatest oxidative potential in the muscle (Essén-Gustavsson and Lindholm, 1985). It has also been shown after racing that the % of type II fibres is positively correlated to ammonia concentrations in muscle and negatively correlated to ATP concentrations (Essén-Gustavsson and Valberg, 1987). This indicates that the AMP-deaminase reaction plays an important role for energy release in type II fibres. In most horses the lowest oxidative capacity is usually seen within type IIB fibres (Essén et al., 1980; Valberg and Essén-Gustavsson, 1987). The oxidative capacity can however markedly vary among horses within type IIB fibres (Valberg et al., 1988). Since type IIB fibres usually only get recruited at maximal speeds the metabolic profile of these fibres and the time when they get recruited are important factors influencing the metabolic responses to exercise. Horses which have many type I fibres and few type IIB fibres, a high oxidative capacity in their fibres and a good capillary supply are able to better maintain high ATP concentrations after a race and they have a lower lactate accumulation in the fibres (Valberg and Essén-Gustavsson, 1987).

Individual fibres have been dissected out from muscle biopsies obtained from Standardbred trotters after racing over 1640 - 2640 m (Essén-Gustavsson et al., 1997). These fibres were typed into being either a type I or a type II fibre. The ATP concentration was shown to markedly vary among these individual fibres with extremely low ATP levels (1-5 mmol/kg) in some fibres and high ATP-levels (40-58 mmol/kg) in other fibres. Mean ATP levels in whole muscle after racing was 18 mmol/kg that is around 70% of resting values. This shows that large variations in ATP levels occur among different fibres and that metabolite analyses on whole muscle must be interpreted with caution. IMP concentrations were also shown to vary among individual fibres after racing and high IMP concentration was associated with low ATP concentration (Essén-Gustavsson et al., 1997). Even if only a few type I fibres could be dissected out, none of these showed as low ATP or as high IMP levels as could be seen in some type II fibres. An earlier study showed that pools of type I fibres had greater ATP levels than pools of type II fibres after racing and the lowest ATP concentration was observed in pools of type IIB fibres (Valberg and Essén-Gustavsson, 1987).

The low ATP and high IMP concentrations found especially in type II fibres after racing indicate that these have been recruited and that rephosphorylation of ADP and AMP cannot match the demand for ATP. An interesting finding is therefore that AMP and ADP concentrations after racing seem to correlate to the placing of the horses in the races (Essén-Gustavsson et al., 1997). The fact that AMP and ADP concentrations were increased in the horses that were not among those placing first indicate that the maximal rate of ATP regeneration in the fibres cannot be maintained and therefore the horses have to lower the speed. This suggests that a loss of adenine nucleotides may be a cause for fatigue development.

Metabolic response to moderate and low intensity exercise

Aerobic metabolism plays an important role for energy release during low intensity exercise. The rise seen in free fatty acids in plasma after exercise of long duration, like endurance rides, indicates an increased rate of lipolysis (Essén-Gustavsson and Jensen-Waern, 2002). Not only glycogen but also triglyceride stores in muscle can decrease after endurance rides (Essén-Gustavsson *et al.*, 1984). In addition, the observed changes in muscle and blood, especially those of the branched chain amino acids, phenylalanine, tyrosine, alanine, glutamine, glutamate and aspartate point to an enhanced amino acid metabolism and protein degradation (Essén-Gustavsson and Jensen-Waern, 2002). Glycogen in muscle is an important substrate for energy release during low and moderate intensity exercise. Many fibres are depleted of glycogen after endurance rides and submaximal treadmill exercise especially type I and type IIA fibres (Essén-Gustavsson *et al.*, 1984; Valberg, 1986; Essén-Gustavsson *et al.*, 1991; Essén-Gustavsson *et al.*, 1999). A decrease in blood glucose can be observed at the end of an endurance ride in some horses suggesting an increased uptake of glucose by the muscle. The blood glucose level will remain constant as long as it can be produced from the glycogen stores in the liver. For optimum performance during low intensity exercise of long duration it is therefore important to have well-filled glycogen stores in muscle and liver.

Anaerobic metabolism becomes more important for energy release as the intensity of submaximal exercise increases. Muscle and blood lactate concentrations are lower after treadmill exercise of low intensity (6-7m/s) to fatigue (50-90 min) compared to after moderate intensity (10 m/s) to fatigue (10-15 min) (Valberg *et al.*, 1989; Essén-Gustavsson *et al.*, 1999). After treadmill exercise to fatigue at both 7 and 10 m/s no marked changes are seen in ATP concentrations measured on whole muscle but IMP concentrations are increased and uric acid concentrations in the blood increase with duration of exercise (Essén-Gustavsson *et al.*, 1999). This indicates that AMP deamination occurs during submaximal exercise and that ATP degradation in some fibres may be one contributing factor for fatigue development. Changes in ATP concentrations are difficult to detect on whole muscle especially if there is a selective depletion of ATP in specific fibres. Further studies on single fibres are therefore needed to investigate ATP and IMP levels within different fibres after submaximal exercise.

Conclusion

Fibre type composition, substrate sources and differences in metabolic properties among fibres and the extent to which fibres are recruited are all factors that influence the metabolic responses of muscle to exercise. Biochemical analyses on whole muscle must be interpreted with caution since large variations in metabolic response to exercise occur among different fibres.

References

Essén, B., Lindholm, A., and Thornton, J. (1980). Histochemical properties of muscle fibre types and enzyme activities in skeletal muscles of Standardbred trotters of different ages. *Equine Veterinary Journal.* 12: 175-180.

Essén- Gustavsson, B., Lindholm, A., McMiken, D., Persson, S.G.B. and Thornton, J. (1983). Skeletal muscle characteristics of young standardbreds in relation to growth and early training. In: *Proceedings from first international conference on Equine Exercise Physiology*, Edited by Snow, Persson and Rose, pp. 200-210.

Essén-Gustavsson, B., Karlström, K. and Lindholm, A. (1984). Fibre types, enzyme activities and substrate utilisation in skeletal muscles of horses competing in endurance rides. *Equine Veterinary Journal.* 16: 197-202.

Essén-Gustavsson, B. and Lindholm, A. (1985). Muscle fibre characteristics of active and inactive Standardbred horses. *Equine Veterinary Journal.* 17: 434-438.

Essén-Gustavsson, B. and Valberg, S. (1987). Blood and muscle ammonia levels in horses during treadmill exercise and after racing. In: *Proceedings from second international conference on Equine Exercise Physiology*, Edited by Gillespie, J.R. and Robinson, N.E., ICEEP Publications, Davis, California, USA, pp. 456-463.

Essén-Gustavsson B, McMiken D, Karlström K, Lindholm A, Persson S, Thornton J. (1989). Muscular adaptation of horses during intensive training and detraining. *Equine Veterinary Journal.* 21: 27-33.

Essén-Gustavsson, B., Blomstrand, E., Karlström, K., Lindholm, A. and Persson, S.G.B. (1991). Influence of diet on substrate metabolism during exercise. In: *Proceedings from third international conference on Equine Exercise Physiology*, Edited by Persson, S.G.B., Lindholm, A. and Jeffcott, L., ICEEP Publications, Davis, California, USA, pp. 288-298.

Essén-Gustavsson, B., Nyman, G. and Wagner, P. (1995). Muscle and blood metabolic responses to intense exercise during acute hypoxia and hyperoxia In: *Proceedings of the fourth international conference on Equine Exercise Physiology,* Edited by N.E. Robinson, *Equine Veterinary Journal (Supplement)* 18: 181-187.

Essén-Gustavsson, B., Roneus. N. and A.R. Pösö. (1997). Metabolic response in skeletal muscle fibres of standardbred trotters after racing. *Comparative Biochemistry and Physiology.* 117 B: 431-436.

Essén-Gustavsson B, Gottlieb-Vedi M. and Lindholm A. (1999). Muscle adenine nucleotide degradation during submaximal treadmill exercise to fatigue. *Equine Veterinary Journal (Supplement);* 30: 298-302.

Essén-Gustavsson B. and Jensen-Waern M. (2002). Effect of an endurance race on muscle amino acids, pro- and macroglycogen and triglycerides. *Equine Veterinary Journal (Supplement)*; 34: 209-213.

Gottlieb M, Essén-Gustavsson B, Lindholm A. and Persson S.G. B. (1988). Circulatory and muscle metabolic responses to draught work compared to increasing trotting velocities. *Equine Veterinary Journal*. 20: 430-434.

Gottlieb M. (1989). Muscle glycogen depletion patterns during draught work in Standardbred horses. *Equine Veterinary Journal*. 21: 110-115.

Gottlieb M, Essén-Gustavsson B. and Skoglund-Wallberg H. (1989). Blood and muscle metabolic responses to draught work of varying intensity and duration in horses. *Research in Veterinary Science*. 47: 102-109.

Karlström, K., Essén-Gustavsson, B., Lindholm, A. and Persson S.G.B. (1991). Capillary supply in relation to muscle metabolic profile and cardiocirculatory parameters. In: *Proceedings from third international conference on Equine Exercise Physiology*. Edited by Persson, S.G.B., Lindholm, A. and Jeffcott, L., ICEEP Publications, Davis, California, USA, pp. 239-244.

Lindholm A. and Piehl K. (1974). Fibre composition, enzyme activity and concentrations of metabolites and electrolytes in muscles of standardbred horses. *Acta Veterinaria Scandinavica*. 15: 287-309.

Lindholm, A., Essén-Gustavsson, B., McMiken, D., Persson, S. and Thornton, J.R. (1983). Muscle histochemistry and biochemistry of thoroughbred horses during growth and training. In: *Proceedings from first international conference on Equine Exercise Physiology*. Edited by Snow, Persson and Rose, pp. 211-217.

Roneus M, Essén-Gustavsson B, Lindholm A. and Persson S.G.B. (1992). Skeletal muscle characteristics in young trained and untrained standardbred trotters. *Equine Veterinary Journal*. 24: 292-294.

Ronéus, M., Essén-Gustavsson, B. and Arnason, T. (1993). Racing performance and longitudinal changes in muscle characteristics in standardbred trotters. *Journal of Equine Veterinary Science*. 13: 355-361.

Schuback K. and Essén-Gustavsson B. (1998). Muscle anaerobic response to a maximal treadmill exercise test in Standardbred trotters. *Equine Veterinary Journal*. 30: 504-510.

Valberg, S., Essén-Gustavsson, B., Lindholm, A. and Persson, S. (1985). Energy metabolism in relation to skeletal muscle fibre properties during treadmill exercise. *Equine Veterinary Journal*. 17: 439-444.

Valberg, S. (1986). Glycogen depletion patterns in the muscle of Standardbred Trotters after exercise of varying intensities and durations. *Equine Veterinary Journal*. 18: 479-484.

Valberg, S. (1987). Metabolic response to racing and fiber properties of skeletal muscle in Standardbred and Thoroughbred horses. *Journal of Equine Veterinary Science*. 7: 6-12.

Valberg, S. and Essén-Gustavsson, B. (1987). Metabolic response to racing determined in pools of type I, IIA and IIB fibres. In: *Proceedings from second international conference on Equine Exercise Physiology*. Edited by Gillespie, J.R. and N. Robinson, ICEEP publication, Davis, California, USA, pp. 290-310.

Valberg, S., Essén-Gustavsson, B. and Skoglund-Wallberg, H. (1988). Oxidative capacity of skeletal muscle fibres in racehorses: histochemical versus biochemical analysis. *Equine Veterinary Journal*. 20: 291-295.

Valberg, S., Essén-Gustavsson, B., Lindholm, A. and Persson, S.G.B. (1989). Blood chemistry and skeletal muscle metabolic responses during and after different speeds and durations of trotting. *Equine Veterinary Journal*. 21: 91-95.

2

Muscle morphology heterogeneity: control, significance for performance and responses to training

J L L Rivero
Laboratory of Muscular Biopathology, Department of Comparative Anatomy and Pathological Anatomy, Faculty of Veterinary Science, University of Cordoba, Cordoba, Spain

Introduction

The skeletal musculature of the horse is highly developed and adapted to match the animal's athletic potential. More than half of a mature horse's body weight comprises skeletal muscle and the total muscle blood flow during maximal exercise represents 78% of total cardiac output. Exercise requires the co-ordinated application of many different body systems under the control of the nervous systems. Metabolites and oxygen reach skeletal muscle fibres via the respiratory, cardiovascular and haematological systems. The muscle fibres produce energy in the form of ATP that, via the contractile machinery, is converted into mechanical work. The structural arrangements of the musculoskeletal system provides the means with which to harness this energy to move the horse's limbs in a characteristic rhythmical pattern that is well established for each gait.

Equine skeletal muscle is considerably heterogeneous and this diversity reflects functional specialisation and is the basis of its adaptive plasticity. Cellular and molecular diversity of equine muscle and the response of this tissue to exercise and training have been studied extensively over the past 30 years. Much of this work has centred around use of the percutaneous needle biopsy technique, a technique originally described for the M. gluteus medius by Lindholm and Piehl (1974). Biochemical and histochemical evaluation of muscle samples have provided invaluable information about different characteristics of equine muscles (see Snow and Valberg 1994 for a review). In recent years, cellular and molecular diversity within equine muscle has also been addressed by means of more objective and refined methods such as immunohistochemistry, gel electrophoresis, enzyme-linked immunosorbent assay and molecular biology techniques. The combined use of these methodologies provides among other things a more objective and accurate way to assess muscle fibre types and their responses to exercise and training. This paper focuses specifically on new information obtained in the past decade from a double perspective. First, assessing muscle fibre type heterogeneity, including its mechanisms of regulation and significance to performance. And second, describing some relevant muscular adaptations to training, as well as possible mechanisms underlying these adaptations and their physiological implications. I will begin with some preliminary considerations about the general muscle structure and function in order to provide an appropriate framework for subsequent discussions.

Muscle structure and function

Locomotor muscles are generally located proximally on the appendicular skeleton, being activated at the end of the stance phase of the stride (Robert *et al.*, 1998). More than 90% of a muscle is made up of myofibres, with the rest consisting of nerves, blood vessels and the fat and connective tissue. The skeletal myofibre is an elongated and multinucleated cell with the nuclei located in a peripheral position. It is mostly taken up by the contractile apparatus that consists of the contractile proteins grouped together as myofibrils. The repeating unit of muscular contraction is called sarcomere. Each sarcomere includes thin and thick filaments. Muscle contraction occurs when, within each sarcomere, thin myofilaments slide over the thick myofilaments. Myosin is the major constituent protein of the thick filaments. Sarcomeric myosins have two heads and a long tail, and consist of two heavy chains and four light chains. The myosin head is the motor domain that contains the ATP-binding site, the actin-binding site and the myofibrillar ATPase enzyme. The major components of the thin filaments are tropomyosin, the troponin complex and the two helical filamentous strands of actin (F-actin), made up of polymerised globular actin monomers (G-actin). Mitochondria are located beneath the sarcolemma and between myofibrils, so ATP produced during oxidative phosphorylation is readily available for the contractile machinery. Intramuscular substrates, such as glycogen and lipids, and numerous other proteins, including myoglobin, glycolytic enzymes and intermediate filaments, are also stored and distributed in the cytoplasm. The sarcoplasmic reticulum is an intracellular membranous system located between and on top of the myofibrils, but has no physical continuity with the external surface membrane (sarcolemma). The motor end plate is a specialised region on the sarcolemma, where the α-motoneuron interdigitates with the sarcolemma. The remainder of the sarcolemma contains a variety of specific membrane proteins that function structurally as channels, pumps and hormone receptors.

Action potential at neuromuscular junctions are conduced into the interior of myofibres via the T-tubules and there activate the calcium release channel with the result that calcium concentration within the cytoplasm increases about a hundred-fold from the resting concentration. Binding of Ca^{2+} to high affinity binding sites of troponin C at the thin filaments causes a conformational change to the troponin-tropomyosin complex. This results in the exposure of the myosin binding sites on F-actin and allows the myosin globular head to attach, as ATP is hydrolysed, thus forming the cross bridge. Force generation and the resulting shortening of the sarcomere are the result of this conformational change of the myosin head. The cross bridge cycle continues while the cytoplasmic Ca^{2+} concentration remains high. Relaxation is achieved when the Ca^{2+} is resequestered within the sarcoplasmic reticulum via the action of the Ca^{2+} ATPase pumps. The force that is generated in the cross bridge cycle is transmitted via the contractile apparatus to intermediate filament proteins of the cytoskeleton. These filaments provide a structural link first to the sarcolemma and then to the extracellular matrix via a group of proteins known as the dystrophin-associated protein complex. Contractile forces are transmitted from each muscle fibre via the extracellular matrix and the connective tissues of tendons ultimately to the bones of the skeleton. ATP replenishment in (predominantly) oxidative fibres requires a readily available source of oxygen that is provided by the protein myoglobin. During exercise oxygen demand rises dramatically and is met by a 20-30 times increases blood flow through the muscle capillary beds. Muscles cannot contract without a biochemical source of energy provided by the cleavage of high-energy phosphate bonds within ATP. At the head of each myosin there is an ATP molecule that becomes hydrolysed and releases energy in a reaction catalysed by the enzyme actomyosin ATPase. Aerobic production of ATP is a relatively slow but highly efficient process, while anaerobic pathways produce energy rapidly, but relatively inefficiently. Although both pathways are generally active during exercise, the relative contribution of each one depends on the nature, intensity and duration of the activity, the muscle's fibre type composition, the availability of oxygen and substrates, and the relative concentrations of intermediary metabolites that may potentially activate or inhibit selected enzymes.

Muscle fibre type heterogeneity

Equine skeletal muscle fibres differ in their morphological, physiological and biochemical properties both within and between muscles. These differences form the basis for the classification of fibre types. Better understanding of the expression patterns of certain proteins within individual fibres has allowed refinement of fibre type classification in recent years. Thus, various myofibrillar and extra-myofibrillar proteins exist as different isoforms encoded by separate genes that are expressed in a myofibre type-specific and co-ordinated manner. Fibre types can best be identified by analysing the specific myosin heavy chain (MyHC) isoform (s) expressed by each fibre, since MyHC composition reflects each fibre's phenotype. Three MyHC isoforms have been characterised by immunoelectrophoresis and immunohistochemistry in adult equine skeletal muscles at the protein level: they are designated as types I, IIA and IIX (Rivero *et al.*, 1999) or IID (Dingboom *et al.*, 1999) (henceforth IIX). This has recently been confirmed at the molecular level (Eizema *et al.*, 2003). The differential distribution of these MyHCs defines three pure fibre types containing a single isoform (types I, IIA and IIX) and two hybrid fibre types co-expressing two isoforms (I+IIA and IIAX). Hybrid IIAX fibres exist in equine locomotor muscles as a significant and stable population (Dingboom *et al.*, 1999; Linnane *et al.*, 1999). No expression of the MyHC-IIB isoform has been demonstrated in the horse (Rivero *et al.*, 1999; Miyata *et al.*, 1999).

Type I fibres have a MyHC isoform that hydrolyses ATP slowly, resulting in a slow cross-bridge cycle, together with a small cross sectional area, a high number of capillaries and a high oxidative capacity. However, their glycolytic capacity and glycogen content are relatively low. Together, these properties make type I fibres highly efficient and economical in producing slow repetitive movements and sustaining isometric force, but not significant power generation. In contrast, type II fibres have MyHC isoforms that create fast cross-bridge cycling and therefore develop force rapidly. Within type II group, type IIX fibres have a maximal velocity of shortening that is three times higher that that of type IIA fibres (Rome *et al.*, 1990). Hence, IIX fibres are adapted for high power outputs for a limited time because they have a low oxidative capacity and limited oxygen availability (as reflected by their large cross sectional area and relatively low capillary supply). Type IIA fibres, however, have a considerable number of both capillaries and mitochondria, and rely on glycolytic and oxidative metabolism; they are therefore able to sustain high power outputs for longer than IIX fibres. Hybrid IIAX fibres are intermediate in their properties (Quiroz-Rothe and Rivero, 2001).

Fibre-type composition varies extensively between muscles and in accordance with their functional requirements. For example, significant components of the forelimb musculature consist of postural type I fibres, while propulsive muscles of the hindlimb contain a high proportion of fast-twitch type II fibres. Variation is also seen between muscles belonging to the same synergic group. For example, most of the triceps muscle mass consists of type II fibres, but the medial head is composed of nearly all type I fibres (Ryan *et al.*, 1992). Significant regional variations in fibre composition within a muscle have also been reported in several horse muscles (Serrano *et al.*, 1996). Most locomotory muscles have greater numbers of oxidative type I and type IIA fibres in the deep portions, and a predominance of glycolytic type IIX in more superficial portions. This compartmentalisation reflects the relationship between structure and function: the deeper regions appear best suited for posture maintenance and low level but longer-duration muscular activity, whereas the more superficial regions are involved with short duration, rapid, propulsive force generation.

Multiple factors, both myogenic and non-myogenic in origin, regulate the expression of proteins that comprise the various muscle-specific organelles in each muscle fibre and, in combination, these factors also regulate the percentage of individual fibre types found within each muscle. During development and maturation these factors change resulting in a significant alteration in protein expression. However, generally, the myogenic lineage from which a muscle fibre develops defines the ultimate fibre type. All fibres during embryonic development express an embryonic MyHC isoform. At birth, some fibres are found to express type I MyHC: these fibres are destined to become mature type I fibres in the adult. Other fibres express a neonatal MyHC isoform that is gradually replaced by either IIA, or IIX, or both MyHCs: these fibres subsequently become type II fibres in the adult. The influence of genetic factors on equine muscle fibre types is clearly illustrated by dramatic variations observed between different breeds of horses (Snow and Guy, 1980) and between separate genealogical bloodlines within the same breed (Rivero et al., 1996). Furthermore, there is a tendency for fibre type ratios (type I: type II) to be inherited (Barrey et al., 1999; Rivero and Barrey, 2001). During growth and maturation, muscle fibres change in their size and histochemical properties: there is a gradual conversion of fast-to-slow phenotype, that is especially pronounced in the first year postpartum (Dingboom et al., 1999) but that may continue until about 6 years of age (Rivero et al., 1993) or older (Gunn, 1995).

In addition to the underlying myogenic lineage, additional factors influence muscle fibre phenotype. Muscle fibres are synciteal (multinucleated), with their myonuclei arranged peripherally throughout the length of the fibre. The volume of cytoplasm associated with a single nucleus is known as the myonuclear domain (Talmadge et al., 1993). Hence, each individual nucleus regulates the expression of proteins within a particular cytoplasmic region. In horses, myonuclear domains of type I fibres are smaller than those of IIX fibres, but similar in size to those of IIA fibres (Quiroz-Rothe and Rivero, 2001). These observations have been related to the different activity patterns of the various fibre types (Roy et al., 1999): the more active type I fibres have a higher rate of both protein synthesis and protein turnover than the less frequently recruited faster fibre types. Neural input also has a significant influence on muscle fibre growth and type (fast/slow, glycolytic/oxidative) through altered regulation of gene expression (Schiaffino et al., 1999). This is convincingly demonstrated by dramatic changes observed in muscles following denervation. Such regulation is particularly evident in slow type I fibres, because their basal neurogenic dependence is significantly greater than faster fibre types (type I > IIA > IIX) (Monti *et al.*, 2001) Additional factors that are known to influence myofibre diversity include hormonal and drug (anabolic steroid)-induced changes (Snow and Valberg, 1994), that may vary depending on the underlying fibre type. Of particular interest, and as discussed in the section that follows, neuromuscular (contractile) activity associated with exercise and training has a significant impact on fibre type adaptation and the expression of fibre-specific protein isoforms.

The basic functional unit of skeletal muscle is the motor unit. A motor unit consists of an a-motor neuron and the skeletal muscle fibres that it innervates. Motor units are commonly classified according to the MyHC profile of their constituent fibres (hence I, IIA, and IIX). This is possible because fibres within a single motor unit show relatively homogenous, although not identical, biochemical and histochemical properties. It appears that motor units are selectively recruited in a specific pattern that changes according to the gait, in addition to the intensity, and duration of exercise. For the maintenance of posture, only type I motor units are recruited. As intensity and duration increase, further motor units are recruited, in the rank order: I → IIA → IIAX → IIX (Valberg. 1996).

Fibre-type composition also varies extensively between individuals and some studies in horses have shown that performance is correlated with selected muscle characteristics (see Snow and Valberg 1994 for a review). Unsurprisingly, endurance capacity is correlated with (1) high percentages of type I and IIA fibres (Rivero *et al.*, 1993), and (2) high activities of oxidative enzymes (Rivero *et al.*, 1995), whereas sprint capacity is correlated with high percentages of type II fibres (Barrey *et al.*, 1999). It is therefore feasible to differentiate the endurance potential of horses based on the fibre type composition of certain muscles (Rivero and Henckel, 1996), although different conclusions have been obtained in Thoroughbreds (Sewell *et al.*, 1994) and trotters (Ronéus and Essén-Gustavsson, 1997). Nevertheless, trotting speed is highly dependent on the ability of muscle to produce energy via anaerobic glycolysis (Ronéus and Essén-Gustavsson, 1997). Interestingly, some myofibre properties are correlated with kinematic profiles. For example, stride length and frequency are positively correlated with both the percentage of IIA fibres (Persson *et al.*, 1991) and fibre size (Rivero and Clayton, 1996). The stance time of the stride is inversely correlated with the percentage of IIX fibres (Ronéus *et al.*, 1995) and fibre diameter (Rivero and Clayton, 1996). Furthermore, some relevant muscular adaptations to training occur with concomitant modifications in the temporal characteristics of the trot (Rivero *et al.*, 2001).

Response to training

Equine skeletal muscle has considerable potential to adapt during training. These long-term adaptations are associated with altered rates and regulation of transcription of specific genes and consequently a change in the amount and/or type of protein expression within muscle fibres. Depending on both the nature (type, frequency, intensity and duration) of the stimulus (exercise training) and on the basal muscle profile, the adaptive response can take the form of: (i) hypertrophy, when myofibres increase in size but otherwise retain their basal structural, physiological and biochemical properties; or (ii) remodelling without hypertrophy, where myofibres do not enlarge but acquire markedly different enzymatic and structural characteristics, often accompanied by changes in the microvasculature; or (iii) a mixed response, i.e., remodelling plus hypertrophy. Overall, muscular adaptations with training have important physiological implications that influence power generation, shortening velocity and resistance to fatigue.

In general, the adaptive response of equine skeletal muscle to training takes the form of remodelling with minimal, if any, hypertrophy. However, specific muscle cell hypertrophy can be stimulated with bursts of muscular activity against high resistance (Heck et al., 1996) and by prolonged stretch beyond normal resting length (Tyler et al., 1998), such as draught-exercise training (Serrano and Rivero, 2000). Six months of conventional jump training also induces a selective hypertrophic growth of type II fibres, with minimal switching between myofibre phenotypes (Rivero and Letelier, 2000). Many other longitudinal studies in horses have also reported significant increases in sizes of type I and/or IIA fibres as an early (less than three months) muscular adaptation to training (reviewed by Snow and Valberg, 1994). This effect is partially explained by a simultaneous fibre type transition in the direction IIX → IIA → I, since fibre size decreases in identical order. In other studies, however, minimal changes or even a reduction in fibre size have been reported (Snow and Valberg, 1994). These observations are hard to reconcile with the prominent increase in muscle mass, especially in the hindquarters, that is generally observed in horses after training. The only explanation for this is a concomitant increase in the number of myofibres (hyperplasia), as previously demonstrated in humans (Sjöström et al., 1991).

Muscle fibre type distribution and MyHC composition are strongly influenced by training. Endurance and strength training in horses have shown an increase of both the IIA:IIX fibre and MyHC ratios and, when training stimulus is high and long enough of type I:II fibre/MyHC ratios (Serrano et al., 2000). A dose-response relationship between the duration (in total) of the training program and the magnitude of induced changes has recently been demonstrated at the molecular level (Serrano *et al.*, 2000). This relationship can be explained more readily in terms of a threshold for the type IIX → IIA transition during the early phase of training, and then a further threshold for the type IIA → type I transition. Sprint training has also been shown to cause increased numbers of type IIA and decreased numbers of IIX fibres (Lovell and Rose, 1991), with corresponding alterations to the respective MyHC content (Rivero *et al.*, 2002). However, a specific decrease of type I fibres has been reported as an early, and probably transitory, response to high-intensity training (Lovell and Rose, 1991; Rivero *et al.*, 2002).

Perhaps the most commonly detected and earliest muscular adaptation to training is an increase in the activity of enzymes of aerobic metabolism. This change is associated with increased mitochondrial and capillary densities. Overall, these changes promote improved oxygen diffusion and more expeditious removal of waste products (such as CO_2). The activities of key anaerobic enzymes either do not change or decrease after training. Training also results in an increase of selected enzymes of purine nucleotide cycle, such as AMP deaminase and creatine kinase, but the concentration of total nucleotide stores is not affected by training (Lovell and Rose, 1991). Training also results in a modest increase in muscle glycogen storage (Ganssen *et al.*, 1999), and increased expression of glucose transporters across the sarcolemma (McCutcheon *et al.*, 2002) and in the transfer of free fatty acids from the vascular to the intracellular compartment via an increase in extracellular albumin concentration (Heilig and Pette, 1988).

Adaptive training responses of skeletal muscles are maintained through 5-6 weeks of inactivity (detraining), but not beyond 12 weeks. A return to sedentary activity levels following a prolonged period of training results in normalisation of expression of MyHC IIX, via a slow-to-fast fibre type transformation in the order I → IIA → IIX (Serrano *et al.*, 2000). These changes occur in parallel with a reversion of the muscle fibre's size, and metabolic and capillary characteristics to pre-training levels (Serrano *et al.*, 2000).

Skeletal muscle responds to increased functional demands by specific quantitative and/or qualitative alterations in gene expression. Persistent elevation of neuromuscular activity (i.e. during training) induces a series of concerted changes in gene expression, evoking either myofibre hypertrophy or myofibre remodelling or both (Williams and Neufeur, 1996). Hypertrophy is characterised by an increase in abundance, per fibre, of most protein constituents. This process involves a general and non-specific augmentation of protein synthesis within cells. But remodelling of myofibre phenotype, with minimal or no hypertrophy, is the typical muscular response to training in the horse. During this kind of adaptation, myofibres undergo a striking reorganisation, with selective activation and repression of many genes. Thus, switching among different myofibrillar isoproteins occur in a graded and sequential manner (Pette, 1998). These changes occur in parallel, but not simultaneously over time, and correspond to the changes in enzymatic profiles, cytosolic proteins, and membrane receptors and transporters.

Recent years have seen significant advances in our understanding of the signalling mechanism by which the information contained in specific action potential patterns is decoded by the transcriptional machinery of myofibre nuclei and mitochondria. Acetylcholine released from motor neurons and other signalling molecules of neural origin bind to cell surface receptors on myofibres and trigger intracellular events that may be linked to altered gene expression, and hence to appropriate modifications. Additional signals are probably derived from contracting myofibres experiencing mechanical stresses that perturb the sarcolemma and extracellular matrix, as well as exerting tension via intermediate filaments on the cytoskeleton, organelles and the nucleus (Milner et al., 1999). Changes in the intracellular concentrations of ions and metabolites during chronic muscle contractions have also been implicated in the activation of signalling pathways (Booth and Baldwin, 1996). The messengers include alterations in calcium concentration, acidosis during anaerobic exercise, the marked reduction in phosphorylation potential of the adenylate system ($[ATP]/[ADP_{free}]$), a depletion of the redox state (NADH/NAD) and hypoxia. Among all these factors, the imbalance between energy requirement and energy supply is perhaps the most important signal triggering an appropriate adjustment in contractile and metabolic protein expression. Other important signalling mechanisms implicated in the motor neuron control of myofibre phenotype and/or growth include: a Ras-mitogen-activated protein kinase, calcineurin and a protein kinase B-dependent and rapamycin-sensitive pathway.

Conclusions

Training-induced changes of skeletal muscle have important implications to the physiological response to exercise. The main consequence of increased muscle mass due to myofibre hypertrophy in response to training is to produce a muscle with a greater peak force capacity, because force output is proportional to total cross-sectional area of the fibre mass recruited. At low speeds, this adaptation has an impact on gait, causing a marked reduction of both stance time and stride duration (Rivero et al., 2001). Additionally, such an adaptation has a significant impact on the performance of show jumpers via enhanced power output from the hindquarters (Rivero and Letelier, 2000). Furthermore, because increased power output results in a greater ability to accelerate and may increase stride length, these training adaptations (strength rather than endurance) may be important for racehorses competing over short distances (Snow and Valberg, 1994). But enhanced power through training comes with the cost of a corresponding decline in aerobic potential, because the increased mass of recruited fibres and concomitant rise in ATP utilisation, occurs simultaneously with a relative inability of oxygen to diffuse into the larger fibres.

Remodelling of myofibre phenotype, with minimal or no hypertrophy, in response to training, produces a muscle that is much more resistant to fatigue, but with an intrinsic decrease in maximal velocity of shortening. The rise in resistance to fatigue corresponds to each myofibre's increased oxidative capacity. The reduction in contractile speed is associated with the switch of muscle fibre types, and the increased expression of slower MyHC and other contractile protein isoforms. In a similar, but reciprocal fashion to that described for strength training, some conventional training programs of young racehorses produce a decrease in the size of type II fibres (Ronéus *et al.*, 1993) and a corresponding decline of both speed and force of contraction (Valberg, 1996). Clearly a balance must be acquired that is set at a levels most appropriate for the intended use of the horses: in general, training programs in racehorses should be aimed at the development of muscle properties that optimise an equilibrium between speed, stamina and strength.

Following endurance training, exercise at submaximal intensities elicits optimal delivery of oxygen and blood-borne substrates, and early activation of oxidative metabolism with lower utilisation rate of endogenous carbohydrates and an increase reliance on fat oxidation as an energy source. Muscle glycogen sparing is underlying the delay in the onset of fatigue during this type of exercise. It seems highly probable that all these metabolic adaptations are largely responsible for the increased endurance in the trained state and the lower propensity for individual muscles to fatigue as measured electromyographically following 8 weeks of aerobic conditioning (Cheung *et al.*, 1998). The increased oxidative capacity, which is observed in skeletal muscle after training, occurs concurrently with increased maximum oxygen uptake (Tyler *et al.*, 1998) and a significant reduction in the net rate of muscle glycogenolysis and anaerobic metabolism (Geor *et al.*, 1999). As a consequence, in the trained state, the speed at which a horse begins to accumulate lactate increases gradually (i.e. there is a delay in the onset of lactate accumulation and ATP depletion). This is accomplished by enhanced muscle buffering capacity and more efficient excitation-contraction coupling. Hence collectively, endurance may be enhanced by a wide variety of related factors that delay the onset of fatigue during anaerobic exercise.

References

Barrey, B., Valette, J.P., Jouglin, M., Blouin, C. and Langlois, B. (1999). Heritability of percentage of fast myosin heavy chains in skeletal muscles and relationship with performance. *Equine Veterinary Journal Supplement* 30: 289-292.

Booth, F.W. and Baldwin, K.M. (1996). Muscle plasticity: energy demand and supply processes. In: *Handbook of physiology*. Edited by Rowel, L.B. and Shepherd, J.T., American Physiological Society, Bethesda, MD, USA, pp. 1075-1123.

Cheung, T.K., Warren, L.K., Lawrence, L.M. and Thompson KN. (1998). Electromyographic activity of the long digital extensor muscle in the exercising Thoroughbred horse. *Equine Veterinary Journal* 30: 251-255

Dingboom, E.G., Dijkstra, G., Enzerink, E., Van Oudheusden, H. C. and Weijs, W.A. (1999). Postnatal muscle fibre composition of the gluteus medius muscle of Dutch warmblood foals: maturation and the influence of exercise. *Equine Veterinary Journal Supplement* 31: 95-100.

Eizema, K., van der Burg, M., Kiri, A., Dingboom, E.G., van Oudheusden, H. Goldspink, G. and Weijs, W.A. (2003). Differential expression of equine myosin heavy-chain mRNA and protein isoforms in a limb muscle. *Journal of Histochemistry and Cytochemistry* 51: 1207-1216.

Gansen, S., Lindner, A., Marx, S., Mosen, H. and Sallmann, H.P. (1999). Effects of conditioning horses with lactate-guided exercise on muscle glycogen content. *Equine Veterinary Journal Supplement* 30: 329-331

Geor, R.J., McCutcheon, L.J. and Shen, H. (1999). Muscular and metabolic responses to moderate-intensity short-term training. *Equine Veterinary Journal Supplement* 30: 311-317

Gunn, H.M. (1995). Relative increase in areas of muscle fibre types in horses during growth. *Equine Veterinary Journal Supplement* 18: 209-213.

Heck, R.W., McKeewer, K.H., Alway, S.E., Auge, W.K., Whitehead, R., Bertone, A.L. and Lombardo, J.A. (1996). Resistance training-induced increases in muscle mass and performance in ponies. *Medicine, Science and Sports Exercise* 28: 877-883

Heilig, A. and Pette, D. (1988). Albumin in rabbit skeletal muscle. Origin, distribution and regulation by contractile activity. *European Journal of Biochemistry* 171: 503-508

Lindholm, A. and Piehl, K. (1974). Fibre composition, enzyme activity and concentrations of metabolites and electrolytes in muscles of standardbred horses. *Acta Veterinaria Scandinavica* 15: 28-309.

Linnane, L., Serrano, A.L. and Rivero, J.L.L. (1999). Distribution of fast myosin heavy chain-based muscle fibres in the gluteus medius of untrained horses: mismatch between antigenic and ATPase determinants. *Journal of Anatomy* 181: 363-372.

Lovell, D.K. and Rose, R.J. (1991). Changes in skeletal muscle composition in response to interval and high intensity training. In: *Equine exercise physiology 3*, Edited by Persson, S.G.B., Lindholm, A. and Jeffcott, L.B., ICEEP Publications, Davis, California, USA, pp. 215-222.

McCutcheon, L.J., Geor, R.J. and Hinchcliff, K.W. (2002). Changes in skeletal muscle GLUT4 content and muscle membrane glucose transport following 6 weeks of exercise training. *Equine Veterinary Journal Supplement* 34: 199-204.

Milner, D.J., Taffet, G.E., Wang, X., Pham, T., Tamura, T., Hartley, C., Gerdes, A.M. and Capetanaki, Y. (1999). The absence of desmin leads to cardiomyocyte hypertrophy and cardiac dilation with compromised systolic function. *Journal of Molecular Cell Cardiology* 31: 2063-76

Miyata, H., Sugiura, T., Kai, M., Hiraga A. and Tokuriki M. (1999). Muscle adaptation of Thoroughbred racehorses trained on a flat or sloped track. *American Journal of Veterinary Research* 60: 1536-1539

Monti, R.J., Roy, R.R. and Edgerton, V.R. (2001). Role of motor unit structure in defining function. *Muscle and Nerve* 24: 848-866.

Persson, S.G.B., Essén-Gustavsson, B. and Lindholm, A. (1991). Energy profile and the locomotory pattern of trotting on an inclined treadmill. In: *Equine exercise physiology 3*, Edited by Persson, S.G.B., Lindholm, A. and Jeffcott, L. B., ICEEP Publications, Davis, California, USA, pp. 231-238.

Pette, D. (1998). Training effects on the contractile apparatus. *Acta Physiological Scandinavia* 162: 367-376

Quiroz-Rothe, E. and Rivero, J.L.L. (2001). Co-ordinated expression of contractile and non-contractile features of control equine muscle fibre types characterised by immunostaining of myosin heavy chains. *Histochemistry and Cell Biology* 116: 299-312.

Robert, C., Valette, J.P. and Denoix, J.M. (1998). Surface electromyographic analysis of the normal horse locomotion: a preliminary report. In: *Proceedings of the Conference on Equine Sport Medicine and Sciences*, Edited by Lindner, A., Wageningen Pers, Wageningen, The Netherlands, pp. 80-85.

Rivero, J.L.L. and Clayton, H. (1996). The potential role of the muscle in kinematic characteristics. *Pferdeheilkunde* 12: 635-640.

Rivero, J.L.L. and Henckel, P. (1996). Muscle biopsy index for discriminating between endurance horses with different performance records. *Research in Veterinary Sciences* 61: 49-54.

Rivero, J.L.L. and Letelier, A.I. (2000). Skeletal muscle profile of show jumpers: physiological and pathological considerations. In: *The elite show jumper, Conference on Equine Sports Medicine and Science 2000*, Edited by Lindner, A., Lensing Druck, Dortmund, Germany, pp. 57-76.

Rivero, J.L.L. and Barrey, E. (2001). Heritabilities and genetic and phenotypic parameters for gluteus medius muscle fibre type composition, fibre size and capillaries in purebred Spanish horses. *Livestock Production Science* 72: 233-241.

Rivero, J.L.L., Galisteo, A.M., Agüera, E. and Miró, F. (1993). Skeletal muscle histochemistry in male and female Andalusian and Arabian horses of different ages. *Research in Veterinary Sciences* 54: 160-169.

Rivero, J.L.L., Serrano, A.L., Henckel, P. and Agüera, E. (1993). Muscle fiber type composition and fiber size in successfully and unsuccessfully endurance-raced horses. *Journal of Applied Physiology* 75: 1758-1766.

Rivero, J.L.L., Serrano, A.L. and Henckel, P. (1995). Activities of selected aerobic and anaerobic enzymes in the gluteus medius muscle of endurance horses with different performance records. *Veterinary Record* 137: 187-192.

Rivero, J.L.L., Valera, M., Serrano, A.S. and Vinuesa, M. (1996). Variability of muscle fibre type composition in a number of genealogical bloodlines in Arabian and Andalusian horses. *Pferdeheilkunde* 12: 661-665.

Rivero, J.L.L., Serrano, A.L., Barrey, E., Valette, J.P. and Jouglin, M. (1999). Analysis of myosin heavy chains at the protein level in horse skeletal muscle. *Journal of Muscle Research and Cell Motility* 20: 211-221.

Rivero, J.L.L., Serrano, A.L., Quiroz-Rothe, E. and Aguilera-Tejero, E. (2001). Co-ordinated changes of kinematics and muscle fibre properties with prolonged endurance training. *Equine Veterinary Journal Supplement* 33: 104-108.

Rivero, J.L.L., Sporleder, H.P., Quiroz-Rothe, E., Vervuert, I., Coenen, M. and Harmeyer, J. (2002). Oral L-carnitine combined with training promotes changes in skeletal muscle. *Equine Veterinary Journal Supplement* 34: 269-274.

Rome, L.C., Sosnicki, A.A. and Gobble, D.O. (1990). Maximum velocity of shortening of three fibre types from horse soleus muscle: implications for scaling with body size. *Journal of Physiology (London)* 431: 173-185.

Ronéus, N. and Essén-Gustavsson, B. (1997). Skeletal muscle characteristics and metabolic response to exercise in young Standardbreds. *American Journal of Veterinary Research* 58: 167-170.

Ronéus, N., Essén-Gustavsson, B., Johnston, C., Drevremo, S. and Persson S. (1995). Lactate response to maximal exercise on the track: relation to muscle characteristics and kinematic variable. *Equine Veterinary Journal Supplement* **18**: 191-194.

Roy, R.R., Monkey, S.R. and Allen D.L. (1999). Modulation of myonuclear number in functionally overloaded and exercised rat plantaris fibers. *Journal of Applied Physiology* **87**: 634-642.

Ryan, J.M., Cobb, C.A. and Hermanson, J.W. (1992). Elbow extensor muscles of the horse: postural and dynamic implications. *Acta Anatomica (Basel)* **144**: 71-79.

Schiaffino, S., Murgia, M., Serrano, A.L. Calabria, E. and Pallafacchina, G. (1999). How is muscle phenotype controlled by nerve activity? *Italian Journal of Neurological Sciences* **20**: 409-412.

Serrano, A.L. and Rivero, J.L.L. (2000). Myosin heavy chain profile of equine gluteus medius muscle following draught-exercise training and detraining. *Journal of Muscle Research and Cell Motility* **21**: 235-245.

Serrano, A.L., Petrie, J.L., Rivero, J.L.L. and Hermanson, J.W. (1996). Myosin isoforms and muscle fiber characteristics in equine gluteus medius muscle. *The Anatomical Record* **244**: 444-451.

Serrano, A.L., Quiroz-Rothe, E. and Rivero, J.L.L. (2000). Early and long-term changes of equine skeletal muscle in response to endurance training and detraining. *Pflügers Archives - European Journal of Physiology* **441**: 263-274.

Sewell, D.A., Harris, R.C. and Marlin, D.J. (1994). Skeletal muscle characteristics in 2 year-old race-trained Thoroughbred horses. *Comparative Biochemistry and Physiology Comparative Physiology* **108**: 87-96.

Sjoström, M., Lexell, J., Eriksson, A. and Taylor, C.C. (1991). Evidence of fibre hyperplasia in human skeletal muscles from healthy young men. *European Journal of Applied Physiology* **62**: 301-308.

Snow, D.H. and Guy, P.S. (1980). Muscle fibre type composition of a number of limb muscles in different types of horses. *Research in Veterinary Sciences* **28**: 137-144.

Snow, D. H. and Valberg, S.J. (1994). Muscle anatomy, physiology and adaptations to exercise and training. In: *Principles and practice of equine sports medicine: the athletic horse*, Edited by Hodgson, D.R. and Rose, R.J., W B Saunders Company, Philadelphia, USA, pp. 145-179.

Talmadge, R.J., Roy, R.R. and Edgerton, V.R. (1993). Muscle fiber types and function. *Current Opinion in Rheumatology* **5**: 695-705.

Tyler, C.M., Golland, L.C. Evans, D.L., Hodgson, D.R. and Rose, R.J. (1998). Skeletal muscle adaptations to prolonged training, overtraining and detraining in horses. *Pflügers Archives - European Journal of Physiology* **436**: 391-397.

Valberg, S.J. (1996). Muscular causes of exercise intolerance in horses.

Veterinary Clinics of North America: Equine Practice **12**: 495-515.

Williams, R.S. and Neufer, P.D. (1996). Regulation of gene expression in skeletal muscle by contractile activity. In: *Handbook of physiology: integration of motor, circulatory, respiratory and metabolic control during exercise*. Edited by Rowell, L.B. and Shepherd, J.T., American Physiological Society, Bethesda, MD, USA, pp. 1124-1150.

3

The mechanics of equine locomotion: a review of progress in our understanding of the mechanical basis of locomotor performance and disease

J F Burn
Department of Anatomy, University of Bristol, Southwell Street, Bristol, BS2 8EJ, UK
J.F.Burn@bris.ac.uk

Introduction

In this chapter we will review how the principles of mechanics can be used to address scientific questions relating to equine performance and disease.

Performance

Performance can be defined in terms of fundamental qualities or at an applied level. Energetic efficiency, speed, acceleration, endurance capacity, stability, and manoeuvrability, are recognised as fundamental selection pressures in the evolution of terrestrial locomotion. These contribute individually or in combination to performance in a particular sport. Whereas the capacity to achieve the highest value of the outcome measure for a particular sport might be used as an overall measure of performance however, it is more useful for the purpose of scientific investigation to consider performance in terms of more fundamental quantities. The overall aim is generally to relate locomotor form (morphology), function (movement), and environmental factors to fundamental measures of performance.

Form

Size (indicated by body mass) has an important effect on the form of the locomotor systems of terrestrial animals. It is clear that horses and animals of similar body mass are ungulates, standing on hooves with a fairly upright limb posture, but the reasons why this architecture is preferable only become apparent if horses are compared to other terrestrial mammals over a wide size range. The fundamental determinants of limb architecture are an animal's ancestry, planetary gravitation, and the properties of the musculoskeletal tissues. Because all animals are made of the same materials (bone for example has similar properties whether from an elephant or a mouse), and gravity is constant, specific changes in limb architecture are necessary to allow animals of different sizes to move in the same way. In addition, theoretical models of locomotion can be used to predict the optimum limb stiffness to allow animals to move in a stable and economic manner. In combination, theoretical models of locomotion and experimental studies can explain the main morphological characteristics of the equine locomotor system in relation to performance requirements.

Function

During locomotion, terrestrial animals move their limbs in characteristic patterns known as gaits. Different gaits are necessary to allow movement at different speeds, and to allow differently shaped animals to travel at the same speed. It has been shown experimentally that the minimum cost of transport is approximately the same for horses travelling at walk, trot, and gallop, and that horses will choose to travel at the most energetically efficient speed within any gait. It is a fundamental consequence of using legs to achieve terrestrial locomotion that fast movement requires an aerial phase during which all the legs are off the ground. This should increase the energy required for locomotion, as it is necessary to work against gravity to lift the body centre of mass. In practice however, horses and other ungulates use elastic energy storage in the digital flexor tendons to increase their energetic efficiency at higher speeds, compensating for an increased vertical movement of the body centre of mass. Recently it has been shown that a similar passive mechanism is used to provide energy for forelimb protraction, and that folding the limb during protraction can occur with little if any energetic input. There is theoretical and experimental evidence to support the hypothesis that the movement exhibited by horses is optimised for energetic economy and efficiency.

Environmental factors that affect performance

The main environmental factors to affect performance are the gradient of the ground surface, the mechanical properties of ground surface, and air resistance. Air resistance has a relatively small effect unless there is a strong following or head wind, in which case the energetic cost of locomotion is decreased or increased respectively. The effect of gradient is also to increase (upward gradient) or decrease (downward gradient) the energetic cost of locomotion. Relatively accurate predictions of the effect of both these factors can be calculated. The mechanical properties of a track surface affect stability, the magnitude of impact forces and energetic cost of locomotion. The coefficient of friction between the hoof and a track surface determines the maximum horizontal force that can be applied between the hoof and the track without slippage. In principle, the coefficient of friction might limit maximum acceleration, deceleration and stability but in practice, surfaces used for equine sports deform to some extent and additional resistance to slippage is provided by the hoof sinking into the surface. The effect of surface deformation on energetic cost of transport is well known, especially in the racing industry where 'going' is measured as a correlate of energy dissipation by the racetrack. A track surface that did not deform would not dissipate mechanical energy but would generate large impact forces on foot placement. The hardness of equestrian sports surfaces is therefore a compromise between preventing large impact forces and maintaining performance by preventing the loss of energy to the surface.

Musculoskeletal disease and injury

Disease or injury causing lameness is of particular significance to horse owners and the equine industry. Scientific research has been aimed at understanding the mechanical factors and events that cause musculoskeletal dysfunction, and developing techniques for diagnosis. Specific goals have been to identify the aetiopathology of musculoskeletal diseases, to identify the optimum mechanical environment for rehabilitation from injury, to develop techniques for early (sub-clinical) diagnosis of musculoskeletal tissue injury and to recognise situations that lead to a high risk of injury. It is difficult to identify a scientific development in any of these areas that has substantially superseded the knowledge accumulated through the practical experience of generations of trainers, riders, and veterinary surgeons. The underlying reason for this is the high complexity of locomotion when analysed at the level of detail necessary to address these questions. Complexity arises from several sources. Firstly, musculoskeletal tissues are, in engineering terms, complex materials due to their anisotropic inhomogeneous composition. In addition, they have the capacity to repair themselves and adapt in response to a mechanical stimulus. This means that the time-scale over which materials are damaged is important in addition to the actual amount of damage caused. An additional layer of complexity is added as these materials are composed into structures within the limbs that are complex in their shape and their interconnections to surrounding structures. Finally, the forces experienced by the limbs during locomotion are complex and difficult to measure. Mechanically induced damage can also occur via two mechanisms, 'single event overload' and fatigue wear. Single event overload occurs as a random event, such as a horse landing awkwardly from a jump, which is difficult to predict or measure. Fatigue damage occurs as the result of the cumulative effect of repeated loading over a large number of strides and is therefore equally difficult to measure. For all these reasons it has proved impossible to model or measure the stress concentrations that occur in the musculoskeletal system with sufficient accuracy to predict the site or extent of mechanically induced damage.

The diagnosis of lameness using objective gait analysis of has been equally unsuccessful in the sense that it is not used widely in clinical practice, and provides no clear advantage over human observation of a moving horse. The reasons for this are less clear. Firstly, there is an obvious disadvantage in that the equipment used to quantify the kinematics and kinetic of equine locomotion is expensive, i.e. upwards of £50,000. This would not however preclude its use by large equine veterinary centres if clear benefits could be identified. A more subtle reason might be that equine morphology prevents diagnosis using gait analysis. Gait analysis is used widely and successfully in clinical human orthopaedics however the joints of the human limb have more degrees of freedom than those of the equine limb. In simple terms this means that humans can limp in more ways than horses, but in mathematical terms is means that there is less diagnostic information available from gait analysis in horses. The reason that our state-of-the-art computerised techniques for quantifying movement are no better than watching a horse trot up might well be because there is nothing more to see than can be seen by eye.

Conclusion

The success in applying mechanics principles and techniques to answer questions concerning locomotion is related inversely to the complexity of the system at the level of detail necessary to address a particular question. Some fundamental questions can be addressed using surprisingly simple models of the locomotor system in which, for example, the body is represented by a single mass, and all legs are represented by a single spring. A reasonably good understanding of equine performance has been obtained using relatively simple models and there is scope for further development in this area in the future. A reasonably good understanding of the role of mechanical factors in the aetiopathology of musculoskeletal disease however, requires a model of locomotor function at a level of detail which appears to be currently outside our capacity for modelling or measurement. It is difficult to predict where future scientific breakthroughs will occur in this area, but it is possible that, for example, the current emphasis on the epidemiology of racing injuries might lead to new approaches to experimental investigation.

4

Breathing and bleeding - the problems of being big and running fast

D Marlin
Head of Physiology, Animal Health Trust, Newmarket, Suffolk and Visiting Professor of Cardiovascular and Respiratory Physiology, University of Bristol

The ability to use oxygen during exercise is determined by a combination of the muscle mitochondrial density, the relative proportion of total body mass consisting of locomotory muscle, the gas exchange surface area and the capacity for conduction of oxygen from the gas exchange surface to the muscle cells, i.e. the heart and circulation. Other factors include: the ability to ventilate the lungs to ensure that a high gradient is maintained between the alveolar oxygen tension and that in the pulmonary capillaries to facilitate diffusion; the capillary density within muscle, the size of the muscle cells and the blood oxygen carrying capacity, which is primarily determined by the total blood volume and haemoglobin concentration.

In absolute terms, the fastest animal on land is the Cheetah, which can reach speeds of up to 120 kmh. However, the Cheetah is a "sprinter" and can only achieve these speeds for distances of around 200 metres (<10 seconds). The maximal capacity to use oxygen during exercise is known as VO_{2max} and shows a good correlation with maximum sustainable (>60-90 seconds) running speed. Whilst the VO_{2max} of the Cheetah has not been reported, it is not likely to be outstandingly high as the short duration and high speed is more indicative of reliance on anaerobic power (ATP, phosphocreatine and lactic acid). Why does the Cheetah need to run so fast? Because its favourite food, gazelles and other antelopes, are athletic and able to achieve speeds of up to 95 km/h. The difference is that the antelopes and gazelles are able to keep this speed up for many minutes because they have an incredibly high capacity to utilise oxygen during exercise – around 300 ml O_2/kg/min for the Pronghorn antelope. To put this in context of man, an average human will have a VO_{2max} around 35 ml O_2/kg/min, whilst that for an elite human distance runner will be of the order of 70-80 ml O_2/kg/min. The highest reported maximal oxygen uptake for a human was ~100 ml O_2/kg/min for the cyclist Miguel Indurain.

Among the larger mammals the horse is clearly an elite athlete in terms of absolute speed, its endurance and also its capacity to use oxygen. For example, the fastest horses are American Quarterhorses which can reach speeds of up to 90 kmh over distances of 200-400 metres. Similar to the cheetah, these horses have a very high anaerobic capacity and maximal oxygen uptakes in the region of 120 ml O_2/ kg/min. This is still high by human standards, but not in comparison to other athletic animal species. Over longer distances, Thoroughbred racehorses can achieve average speeds of 65 kmh over 1000m (or 5 furlongs – the shortest Thoroughbred sprint races) to 60 kmh over 2400 metres (12 furlongs or 1.5 miles) – the English Derby distance. Elite horses running over distances of more than 1600 metres (or 1 mile) and for more than 1.5 minutes have very high maximal oxygen uptakes. Whilst the average Thoroughbred racehorse running over these distances would have a VO_{2max} of the order of 160 ml O_2/kg/ min, we have recorded values as high as 219 ml O_2/kg/min in true elite racehorses – in this case in a champion hurdler. We should also consider that in many cases these feats are achieved whilst carrying an extra 10-15% of its bodyweight in the form of a rider. All the land animals capable of running faster than the horse are smaller in size. For example, whilst the ostrich can run at 90 kmh, it only weighs around 150 kg. Similarly, the pronghorn antelope with a top speed of 95 kmh only weighs around 50kg. So the horse is the fastest of the large mammals.

Why is the horse such a great athlete? In answering this question we must take care to be more specific about breed. In the dog world, breeds like greyhounds are fast, but it would be wrong to say all dogs are fast – we would not describe Bulldogs as athletic, fast or having great endurance. The same is true in the horse world. Ponies only have a VO_{2max} of the order of 100 ml O_2/kg/min. Still high by human standards but poor in comparison to the Thoroughbred. Perhaps the question should be "what makes the Thoroughbred such a great athlete". Firstly, for its bodymass, the size of the heart is relatively large. In an average ability Thoroughbred racehorse the mass of the heart will be of the order of 1.0% of bodyweight. This is high compared to dogs (~0.7%), man (~0.5%) or rodents (~0.4%). Secondly, the horse has the ability to store red blood cells in the spleen. These are released into the circulation at the onset of exercise increasing the capacity of the blood to transport oxygen. There is also a redistribution of fluid and in combination with the splenic release of RBC, there is an increase in total blood volume. Thirdly, the proportion of total bodymass comprised of locomotory muscle is high in the Thoroughbred – around 53%, compared to 40% for man.

Being a large animal and running fast presents a number of physiological and biomechanical challenges not faced by smaller mammals. The physical size of the chest, weight of the lungs, gravity, the problems of inertia and high upper airway resistance, due in part to obligatory nasal breathing and a long trachea, means that the horse is unable to achieve the levels of ventilation achieved by smaller mammals, at least certainly in terms of respiratory frequency. Maximal minute ventilation for a 500kg horse is of the order of 2000 l/min during maximal exercise. However, even at this level of ventilation the horse hypoventilates, as evidenced by increases in arterial carbon dioxide tension. There is an exponential increase in the energetic cost of ventilating the lungs during exercise (referred to as the work of breathing) and beyond around 2000 l/min any increase oxygen consumption achieved by increasing ventilation would be used entirely for work of breathing and thus not result in any increase in muscle VO_2 and hence performance. In fact it is likely that the horse has adopted a one to one relationship between stride and breathing to try and accommodate these biomechanical and energetic constraints. The impact of airway resistance and other biomechanical factors in limiting maximal ventilation have been clearly demonstrated by exercising horses whilst they breathe a mixture of helium and oxygen which has a lower density than air and thus reduces the work of breathing.

The horse, and the racehorse in particular, appears to be prone to a number of disorders of the cardiorespiratory system, such as dynamic upper airway collapse, exercise-induced pulmonary haemorrhage and cardiac valvular disease not seen to the same extent in other athletic animal species. The impact of breeding and selection for speed in relation to these disorders in the athletic horse remains a challenge for performance horse breeders.

Acknowledgements

The author would like to acknowledge the contribution of the following colleagues in this work: Bob Colborne, Bob Schroter, Chris Deaton, David Poole, Howard Ericksson, Casey Kindig, Peter Cashman, Paul McDonough.

5

Causal factors of equine stereotypy

S D McBride[1] and A Hemmings[2]
[1] Institute of Rural Studies, University of Wales, Aberystwyth, SY23 3AL, UK
[2] Royal Agricultural College, Cirencester, Gloucestershire, GL7 6JS, UK

Abstract

The commonality in neural mechanism (transmitter and brain region) underpinning stereotypy and that associated with motivation of goal-directed behaviours lends support to the idea that the former is derived from the latter. Hughes and Duncan postulated this theory behaviourally within their proposed 1988 model of motivation. This review re-interprets the Hughes and Duncan model to neurochemically describe the motivational development of stereotypic behaviour in the horse.

Introduction

The term 'vice' is used to describe abnormal equine behaviour which is usually stereotypic in nature (Ralston, 1982; Schilder, 1986). This descriptive term is considered misleading since it attributes to the horse the ability to make moral decisions as to whether or not to act in an evil manner (Houpt and McDonnell, 1993). It has also resulted in behaviours that are not necessarily stereotypic by definition, being classified as such; common examples are wood-chewing, biting and coprophagia. In accordance with the definition (a behaviour pattern that is repetitive, invariant and has no obvious goal or function), equine stereotypic behaviours are considered here as crib-biting, weaving and box-walking. The prevalence of weaving, crib-biting and box-walking by horses in the UK have been reported as 9.4, 7.5, 3.8% amongst dressage horses, 9.5, 8.3, 3.6%, amongst event horses and 3.9, 3.1, 5.5% amongst endurance horses respectively (McGreevy et al., 1995a). Similar prevalences have been reported in other countries (Prince, 1987; Redbo et al., 1998; Vecchiotti and Galanti, 1986).

Crib-biting or 'cribbing' is an oral stereotypic behaviour that involves the horse gripping onto a fixed object with its incisor teeth, usually at chest level, leaning back onto its hindquarters and contracting the muscles of the neck to bring its head into an arched position. Depending on the individual horse, air may or may not be drawn into the oesophagus, producing a grunting sound; this is known as windsucking (McGreevy et al., 1995b). In some instances, windsucking is performed without the crib-biting action and is termed aerophagia. The fixed, solid object can vary from any surface within the stable to outside posts or gates. There are also anecdotal reports of animals crib-biting on barbed wire or electric fencing when no other appropriate surface was available. Crib-biting leads to extensive wear of the incisor teeth and may result in loss of condition of the horse (Dodman et al., 1987; Sambraus, 1985). Anecdotal reports suggest that crib-biting horses are also prone to spasmodic colic (Fraser and Broom, 1990; Frauenfelder, 1981; Potter, 1980; Sambraus, 1985), and the monetary value of a crib-biting horse is less than that of an equivalent animal free of vices (McBride and Long, 2001). Horse owners regard crib-biting as unsightly and as a potentially 'contagious' problem (McBride and Long, 2001). Thus, crib-biters are often ostracised within horse yards, banished from the premises altogether or attempts made to physically prevent the performance of the behaviour by the use of a crib-strap (a leather strap with two pieces of metal hinged together at the mid-point) or by surgery. The crib-strap is placed around the neck of the horse, and prevents crib-biting by tightening around the pharynx as the horse attempts to arch its neck during the crib-biting sequence. Myectomy or neurectomy describes the two surgical methods (see Fjeldbourg, 1993, for review) employed to prevent the behaviour; the sternomandibularis, omohyoid and sternothyrohyoid muscle groups, or surrounding nerve tissue, are surgically removed from the neck such that the animal cannot arch its neck. Both methods of prevention are not always successful (Brouckaert et al., 2002).

Weaving, by contrast, is a locomotory behaviour using whole body movement; it involves lateral movement of the head and neck from side to side in a rhythmic repetitive manner with alternation of the weight onto the contralateral foreleg, with respect to the position of the head. Weaving has been reported to lead to orthopaedic problems e.g. strained ligaments (Ralston, 1982), loss of condition and poor performance of the animal (Fraser and Broom, 1990). Weaving is usually performed over the stable door and attempts are often made to prevent the behaviour with the use of anti-weave bars. This method is unsuccessful as the animal often steps back from the doorway and performs the behaviour inside the stable.

Box-walking is, as the term suggests, the action of the horse walking around the box (stable) in a circular repetitive fashion. This may result in loss of condition and sometimes performance (Houpt, 1986). Box-walking is rarely prevented due to the impracticalities of immobilising the horse within a stable environment.

Suggested causes and function of equine stereotypic behaviour

The causal factors of equine stereotypic behaviour have been debated for over one hundred years and are still not clearly understood. The following consequences of stabling horses were attributed to the development of the behaviour:

1. crib-biting was related to 'idleness', because use of cereal-based concentrates instead of forage reduced feeding time (Axe, 1901; Friedberger and Frohner, 1905; Goubaux and Barrier, 1892; Holmes, 1839; Moore, 1912; Yare, 1830);
2. lack of social contact (Moore, 1912);
3. the restrictive nature of the stable environment (Moore, 1912).

These factors are still considered causal today (Houpt and McDonnell, 1993; Kiley-Worthington, 1983; Luescher et al., 1991; Waring, 1983), but are now discussed in terms of frustration (Kiley-Worthington, 1983; Kusunose, 1992; Luescher et al., 1991; Waring, 1983), boredom (Kusunose, 1992; Sambraus, 1985; Waring, 1983), or stress (Houpt, 1981; Luescher et al., 1991) of the animal. Three questionnaire-based studies have supported some of the above speculation (Bachmann et al., 2003; McGreevy et al., 1995c; Redbo et al., 1998). McGreevy et al. (1995) showed that the risk of horses developing weaving decreased when levels of forage were above 6.8kg/day (0.42 vs. 1.00; adjusted odds ratio) and that the risk of abnormal behaviour generally (stereotypies and wood-chewing) increased when stable design prevented visual contact with conspecifics (0.6 vs 1.00). Using a logistic regression model, Redbo et al. (1998) demonstrated that the occurrence of stereotypies correlated with the amount of cereal-based concentrate ($p < 0.01$) and forage ($p < 0.001$) given to the horse, as well as the number of horses per trainer ($p < 0.01$). The authors concluded that horses that spent more time eating or, were in greater contact with stable staff, were less likely to perform a stereotypy. Thus, equine stereotypies appear to result from reduced eating time (as a result of eating concentrates), reduced locomotory behaviour and reduced social contact with conspecifics or other species. Similar conclusions have been drawn from other studies (Bachmann et al., 2003; Waters et al., 2002). Two additional factors have also been identified as important in equine stereotypy development and maintenance; the method by which the animal is weaned (box-weaning leads to the greater risk of stereotypy development) (Waters et al., 2002), and the level of visual stimuli presented to the horse (greater levels of stimuli attenuates the behaviour) (Cooper et al., 2000; McAfee et al., 2002).

Once crib-biting becomes established as part of the daily behavioural repertoire of the animal, it is most apparent as a post-prandial behaviour where ingestion of highly palatable feeds causes an immediate crib-biting response (Gillham *et al.*, 1994; Kusunose, 1992; McBride and Cuddeford, 2001). Palatability has thus been discussed as a causal factor. However, palatable feeds are also starch-rich, cereal-based concentrates that reduce the pH of the digestive system, and it has also been speculated that increased acidity in the stomach or large intestine is the main cause of the behaviour, potentially via visceral discomfort (Johnson *et al.*, 1998). The theory is based on the finding that 'grasping-type' behaviour was significantly reduced ($p < 0.05$) when cereal-based concentrates were supplemented with the non-therapeutic antibiotic virginiamycin. Virginiamycin prevented the normal decrease ($p < 0.05$) in caecal pH associated with ingesting concentrates and the authors suggested a link between the behavioural and the physiological effect of the supplement. However, it should be noted that Johnson *et al.* (1998) also observed that the virginiamycin-supplemented group took longer to eat their ration (values not given). Given that eating time is a recognised parameter of palatability (Bellisle, 1989), the study may in fact suggest that antibiotic supplementation reduced the palatability of the feed and in turn the crib-biting response. However, subsequent studies administering other forms of antacid supplement to crib-biting horses have also reported attenuation of the behaviour (Mills and Macleod, 2002; Nicol *et al.*, 2002). It is now being considered that crib-biting is a mechanism whereby the animal is producing additional alkaline saliva to counteract the acidic consequences of ingesting cereal-based concentrates thus reducing the visceral discomfort (Nicol *et al.*, 2002).

Less attention has been given to the causes of weaving and box-walking, although they are regarded as being similar to those identified for crib-biting, i.e. boredom (Fox, 1965; Houpt, 1981; Ralston, 1982), and frustration (Ralston, 1982; Sambraus, 1981) as a result of the stabling. Weaving appears to be elicited by specific cues associated with the provision of a meal (McBride and Cuddeford, 2001; Sambraus and Radtke, 1989). Furthermore, weaving can also be reduced by increasing the level of visual stimuli that the animal is presented with on a daily basis (Cooper *et al.*, 2000; McAfee *et al.*, 2002).

To summarise:

1. equine stereotypies are crib-biting, weaving and box-walking;
2. equine stereotypies are performed by 4-5% of the stabled-horse population;

3. individuals performing these behaviours are often isolated and/ or attempts are made to prevent the behaviour;
4. the cause of equine stereotypic behaviour has not been fully explained, although it has been associated with a reduction in both eating time, locomotory behaviour and social contact and may be related to the weaning experience of the animal.
5. crib-biting becomes established as a post-prandial behaviour where palatability or starch content of the diet appear to be important causal factors of the behaviour;
6. weaving is associated with the cues that indicate provision of food.

Factors associated with stereotypy development in other species

Although 'stress' is often referred to when describing the aetiology of stereotypic behaviour (Dantzer, 1986; Mason, 1991; Rushen *et al.*, 1993; Sambraus, 1996; Wiepkema and Koolhaas, 1993), there are limitations in using this term. Firstly, it is difficult to define (Selye, 1973) and as a result there is variation in how it is interpreted (Rushen, 1995). Secondly, stressors differ both qualitatively (psychological or physical) and quantitatively (e.g. chronic, acute, chronic intermittent) (Dantzer and Mormede, 1983; Friend, 1991; Ladewig *et al.*, 1993), and different individuals respond (physiologically and behaviourally) differently to the same stressors (Ladewig *et al.*, 1993; Mason and Mendl, 1993). As a consequence, not all stressors (events or environments that cause stress) cause stereotypy (e.g. cold, immobilisation and inescapable electric shock (see Robbins, 1990, for review) and not all individuals develop stereotypy in the same environment (Appleby and Lawrence, 1987; Hansen and Damgaard, 1993). However, although 'stress', may be considered too ambiguous a term, specific stressors have been consistently linked to the development of stereotypic behaviour for several species. These are restricted food supply, restricted locomotion and social isolation. For example, restricted food intake reliably causes stereotypic pecking in poultry (Aggrey *et al.*, 1990; Duncan and Wood-Gush, 1972; Savory *et al.*, 1992), whilst for pigs, this together with restriction of locomotion causes stereotypic head-weaving, chain manipulation, bar-biting and sham chewing (Appleby and Lawrence, 1987; Terlouw *et al.*, 1991). Prevention of locomotion causes stereotypic jumping by bank voles (Cooper and Nicol, 1991), whilst caged mink perform stereotypic pacing and rearing (Mason, 1993) and feed-restriction is reported to have a similar effect (Bildsoe *et al.*, 1991). Furthermore, cows tongue rolled when food was restricted (Redbo and Nordblad, 1997; Redbo *et al.*, 1996) or, when they are confined (Redbo, 1993), restricted feeding or social isolation induces stereotypy in sheep (Marsden, 1986), and social isolation has the same effect in dogs (Hubrecht *et al.*, 1992). Thus, it is evident that for several species including the horse, stereotypies will consistently develop in response to similar environmental conditions.

Although this identifies the causal factors of stereotypy, it does not explain why these behaviours develop i.e. it does not describe the underlying motivation of stereotypy. To provide an explanation requires that the occurrence of stereotypy, in response to specific environmental stimuli, be examined in the context of a proposed model of behavioural motivation.

Hughes and Duncan model

The majority of stereotypies are performed either in the context of stimuli that indicate the arrival of food (Conditioned stimuli [CS]) or, the act of feeding itself (Mason and Mendl, 1997). For example, chain-chewing or stereotypic rooting are more prevalent after pigs have eaten (Terlouw *et al.*, 1993), whereas head-weaving is considered to be a pre-feeding activity (Rushen, 1984). Similarly, stereotypic pacing and pecking are pre- and post-meal activities respectively in chickens (Kostal *et al.*, 1992), and weaving and crib-biting are the equine equivalent (McBride and Cuddeford, 2001). The underlying motivation of these feed-related stereotypies may be explained differently when analysed in context of a proposed model of motivation by Hughes and Duncan (1988) (Figure 5.1).

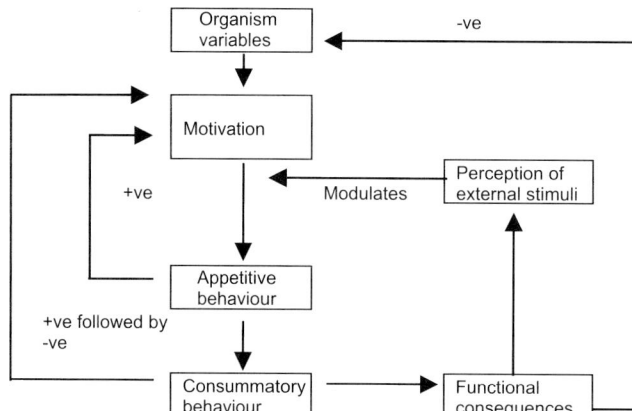

Figure 5.1
Model proposed by Hughes and Duncan (1988), showing the division of behaviour into appetitive and consummatory phases with respective positive and negative feedback mechanisms.

The model is based on the premise that goal-directed behaviours have two distinct components, anticipatory/appetitive and terminal/consummatory. Depending on the state of internal 'organism variables' (e.g. blood glucose levels), motivation to perform consummatory behaviour (e.g. feeding) can result. Consequently, appetitive behaviour takes place (e.g. locomotory action to access food), followed by the consummatory act. The appetitive behaviour has a positive feedback on motivation and is thus self-reinforcing. The act of consummation has a number of effects; 1) functional consequence (e.g. elevated blood glucose), which leads to negative feed-back on organism variables with a subsequent effect on motivation, 2) direct feed-back on motivation, initially positive and then negative, and 3) an effect on perception of the animals environment which again influences the underlying motivation of the behaviour. Hughes and Duncan (1988) argued that stereotypies are appetitive in origin because the restrictive nature of the animal's environments prevents consummatory behaviours from being performed. Lack of consummation and subsequent functional consequence means that no negative feedback on motivation takes place. Appetitive behaviours thus continue and because they are self-reinforcing, the animal gets locked in a positive feed-back loop where the restrictive nature of the environment 'channels' the behaviour into a limited number of discrete acts performed repeatedly. This has been considered to be the basis of stereotypic behaviour (Lawrence and Terlouw, 1993; Rushen et al., 1993). Thus, post-prandial stereotypies in their development and their maintenance are the result of continued eating motivation arising from restriction of food supply; the lack of negative feed-back sustains the motivation to eat resulting in the continued performance of an appetitive behaviour (Lawrence and Terlouw, 1993; Rushen, 1984). Similarly, pre-feeding stereotypies, in response to external stimuli that signal food arrival, develop and maintain as a result of a confined environment which prevents and, thus, sustains the motivation to perform locomotory action to gain food. These pre-prandial stereotypies may become conditioned to the external stimuli, such that with time, the performance of the behaviour becomes emancipated from its original underlying motivation (Mason and Mendl, 1997; Rushen, 1984).

Stereotypies are, however, not always performed in the context of food arrival or feeding. These between-meal or 'facultative' (Rushen, 1984) stereotypies do not differ morphologically from pre- and post-prandial stereotypies, however, they are more difficult to explain in terms of the Hughes and Duncan model (1988) because frustration of appetitive or consummatory behaviours is less apparent. It may be that these behaviours arise from the continued motivation to perform non-feeding consummatory or associated appetitive behaviours e.g. social or sexual behaviour. However, these stereotypies often appear to be in response to either, a lack of or, too much environmental stimuli, both of which may be considered stressful to the animal (for review see Mason, 1991). This has led to the suggestion that stereotypies, in this context, may have a functional capacity in allowing the animal to 'cope' with its environment (for review see Mason, 1991).

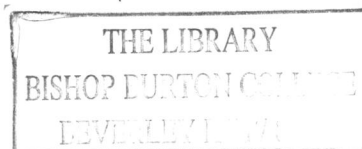

THE LIBRARY
BISHOP BURTON COLLEGE
BEVERLEY HU17 8QG

To summarise, evidence suggests that the majority of stereotypies develop as appetitive behaviours, the response to specific stressors that prevent consummatory (feeding, social contact) and related appetitive (locomotion) behaviours from being performed. These stressors are the result of a confined environment restricting movement and interaction with conspecifics, and also restriction of food supply. Thus, it is reasonable to conclude that, as such, no underlying motivation for stereotypy exists, but rather the behaviour is the manifestation of the motivation to perform other non-obtainable appetitive/consummatory behaviours.

Underlying neurochemical mechanisms of stereotypic behaviour

A substantial body of evidence supports dopamine as the primary, candidate neurotransmitter underlying the development and the maintenance of stereotypic behaviour in several species. This evidence stems from studies where administration of dopamine agonists have been reported to induce stereotypic behaviour (for reviews, see Robbins et al., 1990 and Cooper and Dourish, 1990) and antagonists observed to significantly attenuate these behaviours in several species of animal (Johnson et al., 1994; Kennes et al., 1988; Kostal and Savory, 1994; von Borell and Hurnik, 1991; Willemse et al., 1994). Interestingly, one of the first studies reported (1896) involved administration of the dopamine agonist apomorphine to cause crib-biting behaviour in the horse (Frohner, 1896).

This line of research also identified the striatum as being the primary brain region involved in stereotypic behaviour (see Robbins et al., 1990, for review) and is supported by additional work where injection of a dopamine neurotoxin (6-OHDA) into the striatum region of the brain significantly reduced environmentally-induced stereotypy in rats (Antelman and Szechtman, 1975). The striatum, as part of the basal ganglia, is defined here as the caudate nucleus, putamen, (neostriatum) and the nucleus accumbens (ventral striatum) (Nolte, 2001).

However, other transmitters are also considered to have a role in the development and/or maintenance of the stereotypy response. These include, primarily, opioids and serotonin (5-HT). Table 5.1 presents examples of the neurochemical and behavioural effects of opioid agonist administration specific to the three main sub-types of opioid receptor.

It is evident that administration of opioid agonists will elicit a stereotypic response and this appears to be mediated via dopaminergic pathways within several regions of the brain. These include the neostriatum, the nucleus accumbens, the substantia nigra and the ventral pallidum.

Table 5.1
Summary of the neurochemical and behavioural effects of systemic and local administration of mu, delta and kappa opioid receptor agonists to rodents.

Source	Route/area of administration	Dopamine increase (↑) or decrease (↓)		Behaviour increase (↑) decrease (↓) or no effect(-)	
		Nucleus accumbens	Caudate nucleus	Stereotypic (oral[O], locomotory [L])	Locomotory-non-stereotypic
Mu					
(Mama et al., 1992)	i.v.			↑(O)	↑
(Dichiara and Imperato, 1988)	subcutaneous (s.c.)	↑	↑	-	↑
(Spanagel et al., 1991)	intracerebroventricular (i.c.v)	↑		-	↑
(Morelli et al., 1989)	substantia nigra			↑(O)	→
(Longoni et al., 1991)	nucleus accumbens	→		-	
(Hoffman et al., 1991)	ventral pallidum			↑(L)	
(Longoni et al., 1991)	caudate nucleus		→	-	
(Matsumoto et al., 1988)	substantia nigra			↑(L)	—
Delta					
(Morelli et al., 1989)	substantia.nigra				↑
(Hoffman et al., 1991)	ventral pallidum			↑(L)	
(Spanagel et al., 1990)	i.c.v.	↑			
(Longoni et al., 1991)	nucleus accumbens	↑		↑(O)	
(Meyer and Mclaurin, 1995)	nucleus accumbens			↑(L)	
(Longoni et al., 1991)	caudate nucleus		→		
(Matsumoto et al., 1988)	substantia nigra			↑(L)	—

Table 5.1
Contd.

Source	Route/area of administration	Dopamine increase (↑) or decrease (↓)		Behaviour increase (↑) decrease (↓) or no effect(-)	
		Nucleus accumbens	*Caudate nucleus*	*Stereotypic (oral[O], locomotory [L])*	*Locomotory-non-stereotypic*
Kappa					
(Mama et al., 1992)	i.v.			—	—
(Dichiara and Imperato, 1988)	s.c.	→		—	→
(Spanagel et al., 1990)	i.c.v.	→	→		
(Hoffman et al., 1991)	ventral.pallidum			—	
(Friederich et al., 1987)	substantia nigra			↑(L)	
(Matsumoto et al., 1988)	substantia nigra			↑(L)	

The behavioural effect of opioid agonist administration is also dependent on the type of receptor being stimulated and the location of that receptor within the brain. Evidence suggests that only stimulation of kappa receptors within the substantia nigra region or, mu receptors within the ventral pallidum or substantia nigra regions, will cause stereotypy, whereas stimulation of the ventral pallidum, substantia nigra, neostriatum and nucleus accumbens will elicit stereotypic behaviour via delta receptors. In addition, and supporting the aforementioned conclusions, systemic administration of several opioid antagonists (naloxone, naltrexone and nalmefene) have been reported to attenuate environmentally-induced stereotypic behaviour in several species of animal including the horse (Brown et al., 1987; Cronin et al., 1986; Dodman et al., 1987; Kennes et al., 1988; Kostal and Savory, 1994; McBride and Cuddeford, 2001; Nurnberg et al., 1997; Rushen et al., 1990; Savory et al., 1992; Willemse et al., 1994).

'Serotonin syndrome' describes the behavioural response to systemic administration of any drug that increases the release or availability of serotonin within the central nervous system. Some of these behaviours are stereotypic in nature (e.g. head-weaving and forepaw-treading) as demonstrated when 5-HT agonists were administered to rats (see Curzon, 1990, for review). Furthermore, local infusion of 5-HT into the lateral ventral neostriatum was observed to cause oral stereotypic behaviour in rats which could be attenuated by administering a dopamine antagonist (Yeghiayan and Kelley, 1995).

In summary, research suggests that dopamine is the primary neurotransmitter involved in the underlying neurochemical mechanism of stereotypic behaviour and that the nucleus accumbens, neostriatum, ventral pallidum and substantia nigra are the primary brain regions involved. Other neurotransmitters that modulate dopamine transmission, such as opioids and serotonin, are also considered important in describing the underlying mechanism.

Causal factors of stereotypic behaviour; a neurochemical perspective

Having suggested a neurochemical basis for stereotypic behaviour, the putative causes of these behaviours can now be assessed from a neurochemical perspective.

Causal factors previously implicated in the development of stereotypy were stress generally together with specific stressors associated with feed restriction, restriction of movement and social isolation. The concept of genotypic predisposition to stereotypy development will also be discussed. In addition, factors known to elicit a stereotypy response in the stereotypy-established animal will also be critically assessed from a neurochemical perspective; these include the act of feeding itself, conditioned stimuli associated with the anticipation of meal delivery and stimuli of a general arousing or stressful nature.

Factors associated with the development of stereotypic behaviour

(a)   General and specific stressors

The neurochemical effects of stress are numerous (for review see Anisman and Zacharko, 1990). Examples given in Table 5.2 relate specifically to neurotransmitters and brain regions previously identified to be important in the underlying control of environmentally-induced stereotypy.

Review of the literature, as summarised in Table 5.2, demonstrates that various stressors activate dopamine activity in the neostriatal and nucleus accumbens regions of the brain. Although these are the same brain regions implicated in the control of stereotypic behaviour, a simple relationship between stress-induced dopamine activity and stereotypy does not exist, as stressors not associated with the immediate onset of stereotypy (e.g. restraint and footshock or intermittent immobilisation) have the same neurochemical consequences as those known to elicit a stereotypic response e.g. food restriction (for review see Robbins *et al.*, 1990). However, it should be noted that these stressors do have the ability to enhance (increase in intensity) amphetamine-induced stereotypy suggesting a neurochemical capacity to predispose ('sensitise') the animal to performing the behaviour (Cabib and Bonaventura, 1997; Campbell and Fibiger, 1997; Robbins *et al.*, 1996; Terlouw *et al.*, 1992). These types of stressors are also listed in Table 5.2 and include, isolation rearing and intermittent immobilisation (Cabib *et al.*, 1989; Jones *et al.*, 1992; Kennett *et al.*, 1985; Robbins *et al.*, 1996). The process of 'sensitisation' appears to be dopamine mediated. For example, restraint stress, which enhances apomorphine-induced stereotypy (Cabib *et al.*, 1984), increases D1 and D2 sub-type dopamine receptors in the nucleus accumbens brain region of rodents (Cabib *et al.*, 1998). In addition, mice exhibit stereotypic cage climbing after food restriction (Cabib and Bonaventura, 1997), a stressor that is known to increase dopaminergic transmission of the nucleus accumbens brain region (Cabib *et al.*, 2002). Thus, specific stressors known to induce or enhance stereotypy performance, have a neurochemical consequence associated with increased dopaminergic transmission in the ventral striatum.

Preliminary results suggest that this same mechanism of dopamine mediated predisposition exists for the horse; crib-biting horses have

Table 5.2 The neurochemical effects of stress in rodent species, with specific reference to dopamine, endogenous opioids and serotonin.

Stressor	Is stressor known to cause stereotypy?	Does stressor enhance neuro-transmitter agonist-induced stereotypy?	Brain region					Source
			Striatum	Nucleus accumbens	Frontal cortex	Amygdala	Hippo-campus	
Dopamine								
Low intensity footshock	No				↑			(Inoue et al., 1994)
High intensity footshock	No		↑	↑	↑			(Inoue et al., 1994)
Low conditioned fear stress	No				↑			(Inoue et al., 1994)
High conditioned fear stress	No		↑	↑	↑	↑		(Inoue et al., 1994)}
Social defeat	No		↑	↑	↑			(Tidey and Miczek, 1996)
Cold restraint	No		↑	↑	↑	↑		(Dunn and File, 1983)
Intermittent tail shock	No		↑	↑	↑	↑	↑	(Abercrombie et al., 1989)
Restraint and footshock	No		↑	↑				(Dunn, 1988b)
Isolation rearing	Yes	yes	↑	↑				(Robbins et al., 1996)
Intermittent immobilisation stress	No	yes	↑	↑	↑			(Cabib et al., 1989)
Food restriction	Yes	yes			↑			(Reid et al., 1998)
Food restriction	Yes	yes					↑	(Cabib and Bonaventura, 1997)

Table 5.2
Contd.

Stressor	Is stressor known to cause stereotypy directly?	Does stressor enhance neurotransmitter agonist-induced stereotypy?	Brain region Opioid receptor binding (↑: increase; ↓ decrease)					Source
			Striatum	Nucleus accumbens	Frontal cortex	Amygdala	Hippo-campus	
Opioids								
Food restriction	yes		↑(kappa)					(Wolinsky et al., 1994)
Food restriction	yes		↑(kappa)		↑(kappa)	↓(mu)		(Tsujii et al., 1986)
Water restriction	no		↑(delta)	↑(delta)				(Stein et al., 1992)
Immobilisation stress	no		↑ (mu and delta)					(Zeman et al., 1988)

Table 5.2
Contd.

Stressor	Is stressor known to cause stereotypy directly?	Does stressor enhance neuro-transmitter agonist-induced stereotypy?	Brain region — Inferred serotonin activity as measured by transmitter or metabolite levels or receptor binding (↑: increase; ↓ decrease)					Source
			Striatum	Nucleus accumbens	Frontal cortex	Amygdala	Hippo-campus	
Serotonin								
Low intensity footshock	no							(Inoue et al., 1994)
High intensity footshock	no			↑	↑			(Inoue et al., 1994)
Low conditioned fear stress	no				↑			(Inoue et al., 1994)
High conditioned fear stress	no			↑	↑	↑		(Inoue et al., 1994)
Psychological stress	no			↑	↑	↑		(Kawahara et al., 1993)
Restraint and footshock	no		↑	↑	↑			(Dunn, 1988a)
Chronic immobilisation	no		↑	↑	↑			(Adell et al., 1988)
Isolation rearing	yes			↑				(Jones et al., 1992)
Footshock	no		↑	↑	↑			(Dunn, 1988a)

been reported to have significantly more D1 dopamine receptors in the ventral striatum compared to non-stereotypic animals (McBride and Hemmings, 2001).

(b)   Genotype

Some studies have investigated the genealogy of stereotypic horses and have suggested that crib-biting, weaving and box-walking may have a genetic component (Hosoda, 1950; Vecchiotti and Galanti, 1986). However, these studies have been based on limited data sets and thus firm conclusions about the genetic component of equine stereotypic behaviour are yet to be obtained.

However, in other species, several studies have indicated a genetic basis to stereotypy development. Schwaibold and Pillay (2001) demonstrated that stereotypic offspring were four times more common in stereotypic compared to non-stereotypic females in African striped mice. Schoenecker and Heller (2000) reported stereotypies to be six-seven times more frequent in the offspring of stereotypic bank voles. Similar findings have also been reported in other studies (Hansen, 1993; Odberg, 1986). Research at the neurophysiological level suggests that this genotypic predisposition is again dopamine mediated involving a sensitisation process. For example, mice of the inbred strain DBA/2 (DBA) but not C57BL/6(C57) develop stereotypic behaviour in response to food-restriction (Cabib and Bonaventura, 1997). This stressor is known to enhance striatum dopamine transmission in DBA mice (Cabib et al., 2002). Furthermore, DBA, but not C57, mice respond to immobilisation stress in conjunction with apomorphine injection with increased stereotypy performance (Puglisi-Allegra et al., 1990). This stressor is again linked with upregulation of mesoaccumbens (ventral tegmentum area to nucleus accumbens) dopamine transmission in the former but not the latter strain of mice (Cabib et al., 1998). Thus, it would appear that genotype and stress have an additive effect in predisposing the animal to performing stereotypy and this effect is mediated through changes in dopamine physiology. Given that stereotypy horses also appear to a have altered dopamine physiology, this supports a genetic component to equine stereotypy predisposition (McBride and Hemmings, 2001). It may therefore explain why only some individuals perform stereotypies when kept in the same environment; not all will have the genotype and/or previous history of stress to be sufficiently predisposed/ sensitised to performing the behaviour.

The causal factors of stress and genotype are summarised in the Figure 5.2.

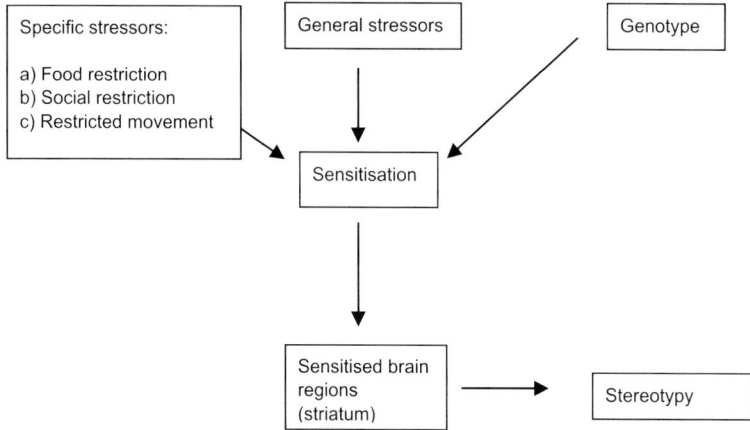

Figure 5.2.
Considered causal
factors of stereotypic
behaviour (model I).

Factors associated with eliciting a stereotypic response in stereotypy-established animals

The neurophysiological consequences of stimuli known to elicit a stereotypic response in established animals (feeding, conditioned stimuli associated with meal delivery and stimuli of a general arousing or stressful nature) indicate a common theme of neural activity; increased transmission of the striatum brain region. Each factor will be discussed separately in this respect.

As previously discussed, feeding stimulates a post-prandial stereotypic response and palatability of feed appears to be important as palatable feeds induce the highest intensity of the oral stereotypy (Gillham et al., 1994; Kusunose, 1992; Yurtman et al., 2002). Although palatability is difficult to define, due to the subjectivity of it's assessment, it is generally agreed to be 'the hedonic response to a taste stimulus based on an integration of taste, the organism's nutritional state and prior association that together, determines acceptability of food ` (Bellisle, 1989; Ossenkopp et al., 1995).

This reward (hedonistic) aspect of palatable food gives an indication of the neural mechanism involved in the underlying neurochemical mechanism of this sensation. Research has shown that reward sensation is associated with dopamine activity within the mesoaccumbens system of the brain, in particular, the nucleus accumbens (for reviews see Wise, 1989, Chen, 1993, and Wrightman and Robinson, 2002) and indeed this is the same mechanism that controls the reward sensation of palatability (in conjunction with endogenous opioid-mediated pathways (Bakshi and Kelley, 1994; Kanarek et al., 1997; Majeed et al., 1986; Martel and Fantino, 1996; Mucha and Iversen, 1986). Thus, given the neurochemical basis of stereotypies, ingestion of a palatable food could have a neurochemical effect that, for the animal in a state of stress-induced sensitisation as previously described, would result in a stereotypic response. This provides an alternative explanation (to the antacid-visceral discomfort theory) for why stereotypic responses are greater when palatable food is offered. However, it does not explain why antacid supplements reduce stereotypic behaviour in the horse unless, as previously discussed, the supplement reduces palatability of the food substrate.

In studies that have examined the neurochemistry of conditioned stimuli associated with the presentation of food, the activation of the nucleus accumbens and the neostriatum again appear, to be intrinsic to the underlying mechanism. For example, Schultz (1986) demonstrated that a light, as a conditioned stimulus to a palatable liquid, caused increased mesostriatal and mesoaccumbens dopaminergic transmission. Using in vivo voltammetry, Phillips et al. (1991) observed dopamine levels to be increased in both the nucleus accumbens and neostriatum during conditioned stimulus indicating food arrival. Furthermore, Blackburn et al. (1989) reported similar results when conditioned stimuli indicating food arrival elicited a greater dopamine response compared to an unexpected meal. Thus, conditioned stimuli associated with food presentation, like eating, would have the same neurochemical and thus behavioural consequences in an animal previously sensitised by stress i.e. elicit stereotypic behaviour.

General stress, as previously discussed, has an overall effect on several neurotransmitter systems including dopamine. Thus, during stressful situations it is reasonable to assume that activation of the striatum region of the brain could induce a stereotypic response.

These eliciting factors in the established animal can be integrated into the previously presented model (Figure 5.3).

Figure 5.3.
Considered causal factors of stereotypic behaviour (model II).

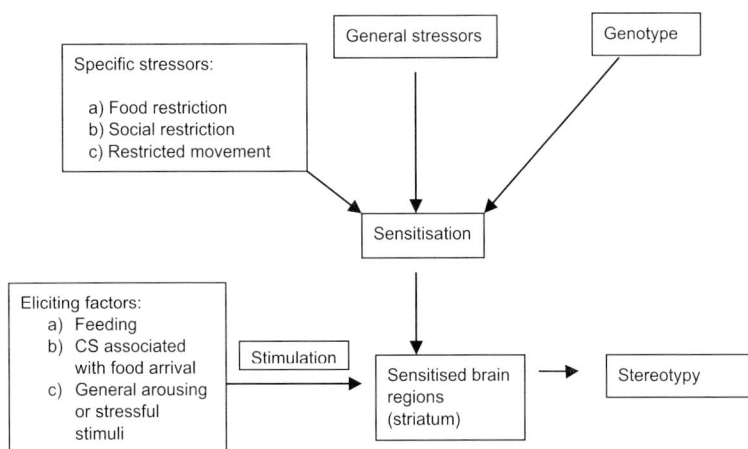

However, although this model provides a neurophysiological link between previously identified causal factors of stereotypic behaviour and the known neural mechanism of the behaviour, it does not adequately explain stereotypic behaviour from a motivational perspective. This can potentially be resolved through neurochemical interpretation of the previously discussed Hughes and Duncan (1988) model.

Neurochemical interpretation of the Hughes and Duncan model

Hughes and Duncan (1988) put forward the idea that stereotypies stem from highly motivated, appetitive behaviours. If one examines the neurochemistry of appetitive behaviours, it is apparent that the dopamine mesolimbic system (in particular the nucleus accumbens region) is the main area of activity (review see Berridge, 1996; Robbins and Everitt, 1996; Berridge and Robinson, 1998; Garris and Rebec, 2002). For example, Kiyatkin and Gratton (1994) demonstrated that accumbens dopamine activity was greatest during the approach and bar-pressing phase of an operant task to obtain food. It follows, therefore, that if an animal has experienced neural sensitisation of this part of the brain (due to stress or genotype or both), the neurochemical consequences of 1) ingestion of a palatable food, 2) conditioned stimuli associated with meal delivery or even, 3) general stressful stimuli, would cause heightened motivation to perform appetitive behaviours (with the aim of reaching a consummatory goal) in these animals. If these behaviours cannot be performed, due to the restrictive nature of the animal's environment, adapted versions of appetitive behaviours (locomotion or non-eating oral behaviour) result i.e. stereotypy. This is supported by the fact that, firstly, consumption of palatable foods (i.e. those foods which cause higher levels of nucleus accumbens stimulation) caused heightened eating motivation (Robert *et al.*, 1997) and the most intense oral stereotypy response (Gillham *et al.*, 1994; Kusunose, 1992; Robert *et al.*, 1997). Secondly, conditioned stimulus (to a rewarding substrate) has been observed to increase sexual behaviour (Mitchell and Stewart, 1990) as well as induce a stereotypic response (Hiroi and White, 1989). Thirdly, with regard to stimuli of a general stressful nature, food–restriction induces psychostimulant self-administration (Carr and Wolinsky, 1993) (for review see Piazza and Le Moal, 1998) and tail-pinch induces eating (Antelman and Szechtman, 1975) and sexual behaviour (Leyton and Stewart, 1996), where both of these stressors are also known to elicit a stereotypic response (Cabib and Bonaventura, 1997; Hawkins *et al.*, 1999). This information is summarised in Figure 5.4.

Figure 5.4 potentially provides a more accurate explanation for why only some stressors elicit stereotypic behaviour. It is now evident that

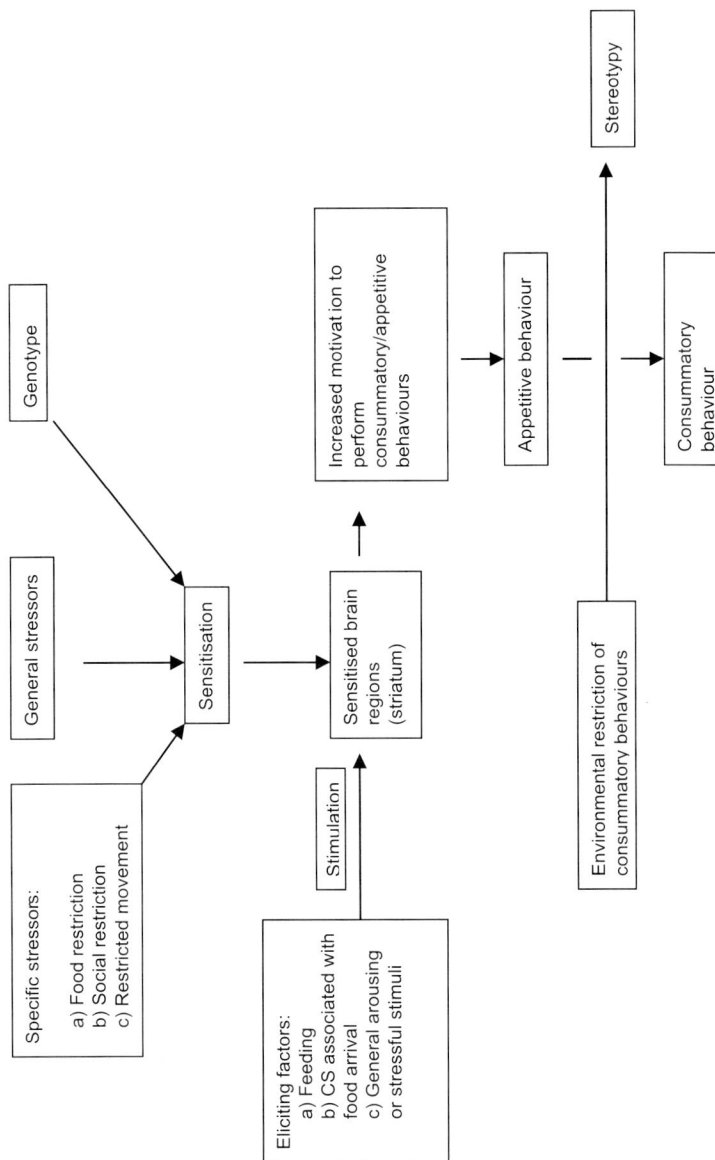

Figure 5.4. Considered causal factors of stereotypic behaviour (model III).

all stressors that have a neural sensitising effect may predispose the animal to perform stereotypy. However, only stressors that prevent the animal from performing consummatory behaviours are those that will actually induce stereotypic behaviour (e.g. confinement or restricted feeding). In this sense and from a causal perspective, equine stereotypies can be described as highly motivated, appetitive behaviours that are the culmination of a neural sensitisation process induced by stress or predisposed by genotype and are performed in an environment where consummatory aspects of the behaviours cannot be satisfied.

References

Abercrombie, E. D., Keefe, K. A., Difrischia, D. S. and Zigmond, M. J. (1989). Differential effect of stress on invivo dopamine release in striatum, nucleus accumbens, and medial frontal-cortex. *Journal of Neurochemistry* 52: 1655-1658.

Adell, A., Garciamarquez, C., Armario, A. and Gelpi, E. (1988). Chronic stress increases serotonin and noradrenaline in rat-brain and sensitizes their responses to a further acute stress. *Journal of Neurochemistry* 50: 1678-1681.

Aggrey, S. E., Kroetzl, H. and Foelsch, D. W. (1990). Behavior of laying hens during induced molting in 3 different production systems. *Applied Animal Behaviour Science* 25: 97-105.

Anisman, H. and Zacharko, R. M. (1990). Multiple neurochemical and behavioral consequences of stressors - implications for depression. *Pharmacology and Therapeutics* 46: 119-136.

Antelman, S. M. and Szechtman, H. (1975). Tail pinch induces eating in sated rats which appear to depend on nigrostriatal dopamine. *Science* 189: 731-733.

Appleby, M. C. and Lawrence, A. B. (1987). Food restriction as a cause of stereotypic behavior in tethered gilts. *Animal Production* 45: 103-110.

Axe, J. W. (1901). Crib-biting and Wind-sucking. *The Veterinary Record* 14: 365-369.

Bachmann, I., Audige, L. and Stauffacher, M. (2003). Risk factors associated with behavioural disorders of crib- biting, weaving and box-walking in Swiss horses. *Equine Veterinary Journal* 35: 158-163.

Bakshi, V. P. and Kelley, A. E. (1994). Sensitization and conditioning of feeding following multiple morphine microinjections into the nucleus-accumbens. *Brain Research* 648: 342-346.

Bellisle, F. (1989). Quantifying palatability in humans. *Annals of the New York Academy of Sciences* 575: 363-375.

Berridge, K. C. (1996). Food reward - brain substrates of wanting and liking. *Neuroscience and Biobehavioral Reviews* 20: 1-25.

Berridge, K. C. and Robinson, T. E. (1998). What is the role of dopamine in reward: hedonic impact, reward learning, or incentive salience? *Brain Research Reviews* 28: 309-369.

Bildsoe, M., Heller, K. E. and Jeppesen, L. L. (1991). Effects of immobility stress and food restriction on stereotypies in low and high stereotyping female ranch mink. *Behavioural Processes* 25: 179-189.

Blackburn, J. R., Phillips, A. G., Jakubovic, A. and Fibiger, H. C. (1989). Dopamine and Preparatory Behavior .2. A Neurochemical Analysis. *Behavioral Neuroscience* 103: 15-23.

Brouckaert, K., Steenhaut, M., Martens, A., Vlaminck, L., Pille, F., Arnaerts, L. and Gasthuys, F. (2002). Windsucking in the horse: results after surgical treatment - A retrospective study (1990-2000). *Vlaams Diergeneeskundig Tijdschrift* 71: 249-255.

Brown, S. A., Crowelldavis, S., Malcolm, T. and Edwards, P. (1987). Naloxone-responsive compulsive tail chasing in a dog. *Journal of The American Veterinary Medical Association* 190: 884-886.

Cabib, S. and Bonaventura, N. (1997). Parallel strain-dependent susceptibility to environmentally- induced stereotypies and stress-induced behavioral sensitization in mice. *Physiology and Behavior* 61: 499-506.

Cabib, S., Puglisi-Allegra, S. and Oliverio, A. (1984). Chronic stress enhances apomorphine-induced stereotyped behavior in mice - involvement of endogenous opioids. *Brain Research* 298: 138-140.

Cabib, S., Oliverio, A. and Puglisi-Allegra, S. (1989). Stress-induced decrease of 3-methoxytyramine in the nucleus accumbens of the mouse is prevented by naltrexone pretreatment. *Life Sciences* 45: 1031-1037.

Cabib, S., Ventura, R. and Puglisi-Allegra, S. (2002). Opposite imbalances between mesocortical and mesoaccumbens dopamine responses to stress by the same genotype depending on living conditions. *Behavioural Brain Research* 129: 179-185.

Cabib, S., Giardino, L., Calza, L., Zanni, M., Mele, A. and Puglisi-Allegra, S. (1998). Stress promotes major changes in dopamine receptor densities within the mesoaccumbens and nigrostriatal systems. *Neuroscience* 84: 193-200.

Campbell, B. A. and Fibiger, H. C. (1997). Potentiation of amphetamine-induced arousal by starvation. *Nature* 233: 424-425.

Carr, K. D. and Wolinsky, T. D. (1993). Chronic food restriction and weight-loss produce opioid facilitation of perifornical hypothalamic self-stimulation. *Brain Research* 607: 141-148.

Chen, J. (1993). Dopaminergic mechanisms and brain reward. *Seminars in the Neurosciences* 5: 315-320.

Cooper, J. J. and Nicol, C. J. (1991). Stereotypic behavior affects environmental preference in bank voles, clethrionomys-glareolus. *Animal Behaviour* 41: 971-977.

Cooper, J. J., McDonald, L. and Mills, D. S. (2000). The effect of increasing visual horizons on stereotypic weaving: implications for the social housing of stabled horses. *Applied Animal Behaviour Science* 69: 67-83.

Cooper, S. J. and Dourish, C. J. (1990). An introduction to the concept of stereotypy and a historical perspective on the role of brain dopamine. In: *Neurobiology of Stereotyped Behaviour*, Edited by Cooper, S. J. and Dourish, C. T., Clarendon Press, Oxford, UK, pp. 1-24.

Cronin, G. M., Wiepkema, P. R. and Vanree, J. M. (1986). Endorphins implicated in stereotypies of tethered sows. *Experientia* 42: 198-199.

Curzon, G. (1990). Stereotyped and other motor responses to 5-hydroxytryptamine receptor activation. In: *Neurobiology of Stereotyped Behaviour*, Edited by Cooper, S. J. and Dourish, C. T., Clarendon Press, Oxford, UK, pp. 142-165

Dantzer, R. (1986). Behavioral, physiological and functional-aspects of stereotyped behavior - a review and a re-interpretation. *Journal of Animal Science* 62: 1776-1786.

Dantzer, R. and Mormede, P. (1983). Stress in farm-animals - a need for reevaluation. *Journal of Animal Science* 57: 6-18.

Dichiara, G. and Imperato, A. (1988). Opposite effects of mu-opiate and kappa-opiate agonists on dopamine release in the nucleus accumbens and in the dorsal caudate of freely moving rats. *Journal of Pharmacology and Experimental Therapeutics* 244: 1067-1080.

Dodman, N. H., Shuster, L., Court, M. H. and Dixon, R. (1987). Investigation into the use of narcotic-antagonists in the treatment of a stereotypic behavior pattern (crib-biting) in the horse. *American Journal of Veterinary Research* 48: 311-319.

Duncan, I. J. H. and Wood-Gush, D. G. M. (1972). Thwarting of feeding behaviour in the domestic fowl. *Animal Behaviour* 20: 444-451.

Dunn, A. J. (1988a). Changes in plasma and brain tryptophan and brain serotonin and 5- hydroxyindoleacetic acid after footshock stress. *Life Sciences* 42: 1847-1853.

Dunn, A. J. (1988b). Stress-related changes in cerebral catecholamine and indoleamine metabolism - lack of effect of adrenalectomy and corticosterone. *Journal of Neurochemistry* 51: 406-412.

Dunn, A. J. and File, S. E. (1983). Cold restraint alters dopamine metabolism in frontal-cortex, nucleus accumbens and neostriatum. *Physiology and Behavior* 31: 511-513.

Fjeldborg, J. (1993). Results of surgical-treatment of cribbing by neurectomy and myectomy. *Equine Practice* 15: 34-36.

Fox, M. W. (1965). Environmental factors influencing stereotyped and allelomimetic behaviour in animals. *Laboratory Animal Care*

15: 363-370.

Fraser, A. F. and Broom, D. M. (1990). Abnormal Behaviour 1: Stereotypies. In: *Farm Animal Behaviour and Welfare*, Bailliere Tindall, UK, pp. 305-317.

Frauenfelder, H. (1981). Treatment of crib-biting - a surgical approach in the standing horse. *Equine Veterinary Journal* 13: 62-63.

Friedberger, F. and Frohner, E. (1905). *Veterinary Pathology Volume 2; Non-infective Diseases of Animals,* London, UK.

Friederich, M. W., Friederich, D. P. and Walker, J. M. (1987). Effects of dynorphin (1-8) on movement - non-opiate effects and structure-activity relationship. *Peptides* 8: 837-840.

Friend, T. H. (1991). Behavioral-aspects of stress. *Journal of Dairy Science* 74: 292-303.

Frohner, E. (1896). Apomorphine. In: *Lehrbuch der Arzneimittellehre,* Stuttgart, Germany, pp. 181.

Garris, P. A. and Rebec, G. V. (2002). Modeling fast dopamine neurotransmission in the nucleus accumbens during behavior. *Behavioural Brain Research* 137: 47-63.

Gillham, S. B., Dodman, N. H., Shuster, L., Kream, R. and Rand, W. (1994). The effect of diet on cribbing behavior and plasma beta- endorphin in horses. *Applied Animal Behaviour Science* 41: 147-153.

Goubaux, A. and Barrier, G. (1892). *The Exterior of the Horse,* London, UK.

Hansen, C. P. B. (1993). Stereotypies in ranch mink - the effect of genes, litter size and neighbors. *Behavioural Processes* 29: 165-178.

Hansen, S. W. and Damgaard, B. M. (1993). Behavioral and adrenocortical coping strategies and the effect on eosinophil leukocyte level and heterophil lymphocyte-ratio in beech marten (martes-foina). *Applied Animal Behaviour Science* 35: 369-388.

Hawkins, M. F., Baumeister, A. A., Larue, R. H., Fountain, L. T., Highsmith, R. W., Jeffries, S. K. and Duke, M. A. (1999). Central GABA activation and behaviors evoked by tail-pinch stress in the rat. *Physiology and Behavior* 67: 705-709.

Hiroi, N. and White, N. M. (1989). Conditioned Stereotypy - Behavioral Specification of the Ucs and Pharmacological Investigation of the Neural Change. *Pharmacology Biochemistry and Behavior* 32: 249-258.

Hoffman, D. C., West, T. E. G. and Wise, R. A. (1991). Ventral pallidal microinjections of receptor-selective opioid agonists produce differential-effects on circling and locomotor- activity in rats. *Brain Research* 550: 205-212.

Holmes, G. (1839). On Crib-biting. *Abstract of the Proceedings of the Veterinary Medical Association* 167-180.

Hosoda, T. (1950). On the heritability of susceptibility to wind-sucking in horses. *Japanese Journal of Zootechnology Science* 21: 25.

Houpt, K. A. (1981). Equine behavior problems in relation to humane management. *International Journal for the Study of Animal Problems* 2 (6): 329-337.

Houpt, K. A. (1986). Stable vices and trailer problems. *Veterinary Clinics of North America-Equine Practice* 2: 623-644.

Houpt, K. A. and McDonnell, S. M. (1993). Equine stereotypies. *Compendium on continuing education for the practicing veterinarian* 15: 1265-1272.

Hubrecht, R. C., Serpell, J. A. and Poole, T. B. (1992). Correlates of pen size and housing conditions on the behavior of kenneled dogs. *Applied Animal Behaviour Science* 34: 365-383.

Hughes, B. O. and Duncan, I. J. H. (1988). The notion of ethological need, models of motivation and animal- welfare. *Animal Behaviour* 36: 1696-1707.

Inoue, T., Tsuchiya, K. and Koyama, T. (1994). Regional changes in dopamine and serotonin activation with various intensity of physical and psychological stress in the rat- brain. *Pharmacology Biochemistry and Behavior* 49: 911-920.

Johnson, K. G., Tyrrell, J., Rowe, J. B. and Pethick, D. W. (1998). Behavioural changes in stabled horses given nontherapeutic levels of virginiamycin. *Equine Veterinary Journal* 30: 139-143.

Johnson, R. W., Vonborell, E. H., Anderson, L. L., Kojic, L. D. and Cunnick, J. E. (1994). Intracerebroventricular injection of corticotropin-releasing hormone in the pig - acute effects on behavior, adrenocorticotropin secretion, and immune suppression. *Endocrinology* 135: 642-648.

Jones, G. H., Hernandez, T. D., Kendall, D. A., Marsden, C. A. and Robbins, T. W. (1992). Dopaminergic and serotonergic function following isolation rearing in rats - study of behavioral-responses and postmortem and invivo neurochemistry. *Pharmacology Biochemistry and Behavior* 43: 17-35.

Kanarek, R. B., Mathes, W. F., Heisler, L. K., Lima, R. P. and Monfared, L. S. (1997). Prior exposure to palatable solutions enhances the effects of naltrexone on food intake in rats. *Pharmacology Biochemistry and Behavior* 57: 377-381.

Kawahara, H., Yoshida, M., Yokoo, H., Nishi, M. and Tanaka, M. (1993). Psychological stress increases serotonin release in the rat amygdala and prefrontal cortex assessed by in-vivo microdialysis. *Neuroscience Letters* 162: 81-84.

Kennes, D., Odberg, F. O., Bouquet, Y. and De Rycke, P. H. (1988). Changes in naloxone and haloperidol effects during the development of captivity-induced jumping stereotypy in bank voles. *European Journal of Pharmacology* 153: 19-24.

Kennett, G. A., Dickinson, S. L. and Curzon, G. (1985). Enhancement of some 5-ht-dependent behavioral-responses following repeated immobilization in rats. *Brain Research* 330: 253-263.

Kiley-Worthington, M. (1983). Stereotypes in horses. *Equine Practice* 5 (1): 34-40.

Kiyatkin, E. A. and Gratton, A. (1994). Electrochemical Monitoring of Extracellular Dopamine in Nucleus-Accumbens of Rats Lever-Pressing for Food. *Brain Research* 652: 225-234.

Kostal, L. and Savory, C. J. (1994). Influence of pharmacological manipulation of dopamine and opioid receptor subtypes on stereotyped behavior of restricted-fed fowls. *Pharmacology Biochemistry and Behavior* 48: 241-252.

Kostal, L., Savory, C. J. and Hughes, B. O. (1992). Diurnal and individual variation in behaviour of restricted fed broiler breeders. *Applied Animal Behaviour Science* 32: 361-374.

Kusunose, R. (1992). Diurnal pattern of cribbing in stabled horses. *Japanese Journal of Equine Science* 3 (2): 173-176.

Ladewig, J., de Passille, A. M. B., Rushen, J., Terlouw, E. M. C. and von Borell, E. (1993). Stress and the physiological correlates of stereotypic behaviour. In: *Stereotypic Animal Behaviour: fundamentals and applications to welfare*, Edited by Lawrence, A. B. and Rushen, J., CAB International, Wallingford, UK, pp. 97-118.

Lawrence, A. B. and Terlouw, E. M. C. (1993). A review of behavioral-factors involved in the development and continued performance of stereotypic behaviors in pigs. *Journal of Animal Science* 71: 2815-2825.

Leyton, M. and Stewart, J. (1996). Acute and repeated activation of male sexual-behavior by tail pinch - opioid and dopaminergic mechanisms. *Physiology and Behavior* 60: 77-85.

Longoni, R., Spina, L., Mulas, A., Carboni, E., Garau, L., Melchiorri, P. and Dichiara, G. (1991). (d-ala2)deltorphin-ii - d1-dependent stereotypies and stimulation of dopamine release in the nucleus-accumbens. *Journal of Neuroscience* 11: 1565-1576.

Luescher, U. A., Mckeown, D. B. and Halip, J. (1991). Reviewing the causes of obsessive-compulsive disorders in horses. *Veterinary Medicine* 86: 527-530.

Majeed, N. H., Przewlocka, B., Wedzony, K. and Przewlocki, R. (1986). Stimulation of food-intake following opioid microinjection into the nucleus accumbens septi in rats. *Peptides* 7: 711-716.

Mama, K. R., Pascoe, P. J. and Steffey, E. P. (1992). Evaluation of the interaction of mu and kappa-opioid agonists on locomotor behavior in the horse. *Canadian Journal of Veterinary Research* 57: 106-109.

Martel, P. and Fantino, M. (1996). Mesolimbic dopaminergic system activity as a function of food reward - a microdialysis study. *Pharmacology Biochemistry and Behavior* 53: 221-226.

Mason, G. and Mendl, M. (1997). Do the stereotypies of pigs, chickens and mink reflect adaptive species differences in the control of foraging? *Applied Animal Behaviour Science* 53: 45-58.

Mason, G. J. (1991). Stereotypies- a critical review. *Animal Behaviour* 41: 1015-1037.

Mason, G. J. (1993). Age and context affect the stereotypies of caged mink. *Behaviour* 127: 191-229.

Mason, G. J. and Mendl, M. T. (1993). Why is there no easy way of measuring animal welfare? *Animal Welfare* 2: 301-319.

Matsumoto, R. R., Brinsfield, K. H., Patrick, R. L. and Walker, J. M. (1988). Rotational behavior mediated by dopaminergic and nondopaminergic mechanisms after intranigral microinjection of specific mu-opioid, delta-opioid and kappa-opioid agonists. *Journal of Pharmacology and Experimental Therapeutics* 246: 196-203.

McAfee, L. M., Mills, D. S. and Cooper, J. J. (2002). The use of mirrors for the control of stereotypic weaving behaviour in the stabled horse. *Applied Animal Behaviour Science* 78: 159-173.

McBride, S. D. and Long, L. (2001). The perception and subsequent management of equine stereotypic behaviour by horse owners; implications for animal welfare. *The Veterinary Record* 148: 799-802.

McBride, S. D. and Cuddeford, D. (2001). The putative welfare-reducing effects of preventing equine stereotypic behaviour. *Animal Welfare* 10: 173-189.

McBride, S. D. and Hemmings, A. (2001). Striatum D1 dopamine receptors are significantly higher in horses perfroming stereotypic behaviour. *Research in Veterinary Science.* 70: 12.

McGreevy, P. D., French, N. P. and Nicol, C. J. (1995a). The prevalence of abnormal behaviors in dressage, eventing and endurance horses in relation to stabling. *The Veterinary Record* 137: 36-37.

McGreevy, P. D., Richardson, J. D., Nicol, C. J. and Lane, J. G. (1995b). Radiographic and endoscopic study of horses performing an oral based stereotypy. *Equine Veterinary Journal* 27: 92-95.

McGreevy, P. D., Cripps, P. J., French, N. P., Green, L. E. and Nicol, C. J. (1995c). Management factors associated with stereotypic and redirected behaviour in the thoroughbred horse. *Equine Veterinary Journal* 27: 86-91.

Meyer, M. E. and Mclaurin, B. I. (1995). Intraaccumbens delta(1)-opioid agonist, pcl-dpdpe, differentially affects patterns of locomotor-activity. *Pharmacology Biochemistry and Behavior* 51: 359-362.

Mills, D. S. and Macleod, C. A. (2002). The response of crib-biting and windsucking in horses to dietary supplementation with an antacid mixture. *Ippologia* 13: 33-41.

Mitchell, J. B. and Stewart, J. (1990). Facilitation of sexual behaviors in the male-rat in the presence of stimuli previously paired with systemic injections of morphine. *Pharmacology Biochemistry and Behavior* 35: 367-372.

Moore, J. (1912). Crib-biting and Wind-sucking. *The Veterinary Record* **24**: 806-814.

Morelli, M., Fenu, S. and Dichiara, G. (1989). Substantia nigra as a site of origin of dopamine-dependent motor syndromes induced by stimulation of mu-opioid and delta-opioid receptors. *Brain Research* **487**: 120-130.

Mucha, R. F. and Iversen, S. D. (1986). Increased food-intake after opioid microinjections into nucleus accumbens and ventral tegmental area of rat. *Brain Research* **397**: 214-224.

Nicol, C. J., Davidson, H. P. D., Harris, P. A., Waters, A. J. and Wilson, A. D. (2002). Study of crib-biting and gastric inflammation and ulceration in young horses. *The Veterinary Record* **151**: 658-+.

Nolte, J. (2001). Organization of the brainstem. In: *The Human Brain; an introduction to its functional anatomy.* Edited by Nolte, J., Mosby, St Louis, USA, pp. 254-282.

Nurnberg, H. G., Keith, S. J. and Paxton, D. M. (1997). Consideration of the relevance of ethological animal models for human repetitive behavioral spectrum disorders. *Biological Psychiatry* **41**: 226-229.

Odberg, F. O. (1986). The Jumping Stereotypy in the Bank Vole (Clethrionomys- Glareolus). *Biology of Behaviour* **11**: 130-143.

Ossenkopp, K. P., Parker, L. A. and Spector, A. C. (1995). Behavioral, neural, and pharmacological aspects of palatability: An introduction to the symposium. *Neuroscience and Biobehavioral Reviews* **19**: 87-88.

Phillips, A. G., Pfaus, J. G. and Blaha, C. D. (1991). Dopamine and motivated behavior: Insights provided by in vivo analyses. In: *The mesolimbic dopmine system: from motivation to action.* Edited by Willner, P. J. and Scheel-Kruger, J., Wiley, Chichester, UK, pp. 199-224.

Piazza, P. V. and Le Moal, M. (1998). The role of stress in drug self-administration. *Trends In Pharmacological Sciences* **19**: 67-74.

Potter, G. D. (1980). Behavioral problems. *Equine Veterinary Data* 194-195.

Prince, D. (1987). Stable Vices. In: *Behaviour problems in horses,* Edited by McBane, S., David & Charles, Newton Abbot, UK, pp. 115-122.

Puglisi-Allegra, S., Kempf, E. and Cabib, S. (1990). Role of genotype in the adaptation of the brain dopamine system to stress. *Neuroscience and Biobehavioral Reviews* **14**: 523-528.

Ralston, S. L. (1982). Common behavioral problems of horses. *Compendium on continuing education for the practicing veterinarian* **4**: S152-S159.

Redbo, I. (1993). Stereotypies and cortisol secretion in heifers subjected to tethering. *Applied Animal Behaviour Science* **38**: 213-225.

Redbo, I. and Nordblad, A. (1997). Stereotypies in heifers are affected by feeding regime. *Applied Animal Behaviour Science* 53: 193-202.

Redbo, I., Emanuelson, M., Lundberg, K. and Oredsson, N. (1996). Feeding level and oral stereotypies in dairy-cows. *Animal Science* 62: 199-206.

Redbo, I., Redbo-Torstensson, P., Odberg, F. O., Hedendahl, A. and Holm, J. (1998). Factors affecting behavioural disturbances in race-horses. *Animal Science* 66: 475-481.

Reid, M. S., Ho, L. B., Tolliver, B. K., Wolkowitz, O. M. and Berger, S. P. (1998). Partial reversal of stress-induced behavioral sensitization to amphetamine following metyrapone treatment. *Brain Research* 783: 133-142.

Robbins, T. W. and Everitt, B. J. (1996). Neurobehavioral mechanisms of reward and motivation. *Current Opinion In Neurobiology* 6: 228-236.

Robbins, T. W., Jones, G. H. and Wilkinson, L. S. (1996). Behavioral and neurochemical effects of early social deprivation in the rat. *Journal of Psychopharmacology* 10: 39-47.

Robbins, T. W., Mittleman, G., OBrien, J. and Winn, P. (1990). The neuropsychological significance of stereotypy induced by stimulant drugs. In: *Neurobiology of stereotyped behaviour*, Edited by Cooper, S. J. and Dourish, C. T., Clarendon Press, Oxford, UK, pp. 25-63.

Robert, S., Rushen, J. and Farmer, C. (1997). Both energy content and bulk of food affect stereotypic behaviour, heart rate and feeding motivation of female pigs. *Applied Animal Behaviour Science* 54: 161-171.

Rushen, J. (1984). Stereotyped behavior, adjunctive drinking and the feeding periods of tethered sows. *Animal Behaviour* 32: 1059-1067.

Rushen, J., Depassille, A. M. B. and Schouten, W. (1990). Stereotypic behavior, endogenous opioids, and postfeeding hypoalgesia in pigs. *Physiology and Behavior* 48: 91-96.

Rushen, J., Lawrence, A. B. and Terlouw, E. M. C. (1993). The motivational basis of stereotypies. In: *Stereotypic animal behaviour: fundamentals and applications to welfare*, Edited by Lawrence, A. B. and Rushen, J., CAB International, Wallingford, UK, pp. 41-64.

Sambraus, H. H. (1981). Abnormal-behavior as an indication of immaterial suffering. *International Journal for the Study of Animal Problems* 2: 245-248.

Sambraus, H. H. (1985). Mouth-Based Anomalous Syndromes. In: *Ethology of farm animals*, Edited by Fraser, A. F., Elsevier, The Netherlands, pp. 391-422.

Sambraus, H. H. (1996). Stereotypies. In: *Ethology of farm animals*, Edited by Fraser, A. F., Elsevier, The Netherlands, pp. 431-441.

Sambraus, H. H. and Radtke, K. (1989). Study with respect to the weaving of horses. *Deutsche Tierartliche Wochenschrift* 96: 248-255.

Savory, C. J., Seawright, E. and Watson, A. (1992). Stereotyped behavior in broiler breeders in relation to husbandry and opioid receptor blockade. *Applied Animal Behaviour Science* 32: 349-360.

Schilder, M. B. H. (1986). Vices in stabled horses with special reference to crib biting. *Applied Animal Behaviour Science* 15: 89-90.

Schoenecker, B. and Heller, K. E. (2000). Indication of a genetic basis of stereotypies in laboratory- bred bank voles (Clethrionomys glareolus). *Applied Animal Behaviour Science* 68: 339-347.

Schultz, W. (1986). Responses of Midbrain Dopamine Neurons to Behavioral Trigger Stimuli in the Monkey. *Journal of Neurophysiology* 56: 1439-1461.

Schwaibold, U. and Pillay, N. (2001). Stereotypic behaviour is genetically transmitted in the African striped mouse Rhabdomys pumilio. *Applied Animal Behaviour Science* 74: 273-280.

Selye, S. (1973). The evolution of the stress concept. *American Scientist* 61: 692-699.

Spanagel, R., Herz, A. and Shippenberg, T. S. (1990). The effects of opioid-peptides on dopamine release in the nucleus- accumbens - an invivo microdialysis study. *Journal of Neurochemistry* 55: 1734-1740.

Spanagel, R., Herz, A., Balskubik, R. and Shippenberg, T. S. (1991). Beta-endorphin-induced locomotor stimulation and reinforcement are associated with an increase in dopamine release in the nucleus- accumbens. *Psychopharmacology* 104: 51-56.

Stein, E. A., Hiller, J. M. and Simon, E. J. (1992). Effects of stress on opioid receptor-binding in the rat central- nervous-system. *Neuroscience* 51: 683-690.

Terlouw, E. M. C., Lawrence, A. B. and Illius, A. W. (1991). Influences of feeding level and physical restriction on development of stereotypies in sows. *Animal Behaviour* 42: 981-991.

Terlouw, E. M. C., Wiersma, A., Lawrence, A. B. and Macleod, H. A. (1993). Ingestion of food facilitates the performance of stereotypies in sows. *Animal Behaviour* 46: 939-950.

Terlouw, E. M. C., Derosa, G., Lawrence, A. B., Illius, A. W. and Ladewig, J. (1992). Behavioral-responses to amphetamine and apomorphine in pigs. *Pharmacology Biochemistry and Behavior* 43: 329-340.

Tidey, J. W. and Miczek, K. A. (1996). Social defeat stress selectively alters mesocorticolimbic dopamine release - an in-vivo microdialysis study. *Brain Research* 721: 140-149.

Tsujii, S., Nakai, Y., Koh, T., Takahashi, H., Usui, T., Ikeda, H., Matsuo, T. and Imura, H. (1986). Effect of food-deprivation on opioid receptor-binding in the brain of lean and fatty zucker rats. *Brain Research* **399**: 200-203.

Vecchiotti, G. G. and Galanti, R. (1986). Evidence of heredity of cribbing, weaving and stall-walking in Thoroughbred horses. *Livestock Production Science* **14**: 91-95.

von Borell, E. and Hurnik, J. F. (1991). The effect of haloperidol on the performance of stereotyped behavior in sows. *Life Sciences* **49**: 309-314.

Waring, G. H. (1983). *Horse Behavior,* Noyes Publications, Park Ridge.

Waters, A. J., Nicol, C. J. and French, N. P. (2002). Factors influencing the development of stereotypic and redirected behaviours in young horses: findings of a four year prospective epidemiological study. *Equine Veterinary Journal* **34**: 572-579.

Wiepkema, P. R. and Koolhaas, J. M. (1993). Stress and animal welfare. *Animal Welfare* **2**: 195-218.

Wightman, R. M. and Robinson, D. L. (2002). Transient changes in mesolimbic dopamine and their association with 'reward'. *Journal of Neurochemistry* **82**: 721-735.

Willemse, T., Mudde, M., Josephy, M. and Spruijt, B. M. (1994). The effect of haloperidol and naloxone on excessive grooming behavior of cats. *European Neuropsychopharmacology* **4**: 39-45.

Wise, R. A. and Rompre, P. P. (1989). Brain Dopamine and Reward. *Annual Review of Psychology* **40**: 191-225.

Wolinsky, T. D., Carr, K. D., Hiller, J. M. and Simon, E. J. (1994). Effects of chronic food restriction on mu-opioid and kappa-opioid binding in rat forebrain - a quantitative autoradiographic study. *Brain Research* **656**: 274-280.

Yare, T. R. (1830). On the Vicious habits and Propensities of Horses. *The Sporting Magazine* **1** (2nd edition): 208-218.

Yeghiayan, S. K. and Kelley, A. E. (1995). Serotonergic stimulation of the ventrolateral striatum induces orofacial stereotypy. *Pharmacology Biochemistry and Behavior* **52**: 493-501.

Yurtman, I. Y., Savas, T., Karaagac, F. and Coskuntuna, L. (2002). Effects of daily protein intake levels on the oral stereotypic behaviours in energy restricted lambs. *Applied Animal Behaviour Science* **77**: 77-88.

Zeman, P., Alexandrova, M. and Kvetnansky, R. (1988). Opiod-mu and opiod-delta and dopamine receptor number changes in rat striatum during stress. *Endocrinologia Experimentalis* **22**: 59-66.

6

The putative reward function of equine stereotypic behaviour

A J Hemmings [1], S D McBride[2] and N C Smith[3]
[1]*Royal Agricultural College, Cirencester, Gloucestershire, GL7 6JS, UK*
[2]*Institute of Rural Studies, University of Wales, Aberystwyth, SY23 3AL, UK*
[3]*Abingdon and Witney College, Holloway Road, Witney, Oxfordshire OX28 6NE, UK*

Abstract

A review of physiology and behaviour-based studies on the function of stereotypic behaviour indicates contradiction and inconsistency within the literature. By considering appropriate neurochemical data alongside an existing model of motivation (Hughes and Duncan, 1988), a greater understanding of the function of stereotypy may be gained. The Hughes and Duncan model (1988), described stereotypies as highly motivated appetitive behaviours performed repeatedly in an environment where consummatory goals are denied. Moreover, appetitive behaviours activate reward circuitry such as the ventral tegmental area and nucleus accumbens and are thus considered to have a reward value associated with their performance (Carr, 2002; Jones et al., 1990). Stress induced sensitisation of reward circuitry may result in appetitive 'stereotypies' having increased reward value, perhaps becoming consummatory in their own right. In such a scenario, stereotypic behaviour could function as a coping tool, allowing the animal to counter the effects of an aversive environment.

Introduction

According to the current definition (Commission of the European Communities, 1983), behavioural stereotypy serves no apparent purpose, and instead may be attributed to factors such as neural dysfunction or the repetitive deployment of an innate behavioural response (Garner and Mason, 2002). However, due to the large impact on time budget, and the apparent high motivation towards performance, it has been suggested that stereotypy may have some form of functional consequence for the animal (Mason, 1991); in particular stereotypies may function as a coping mechanism in response to a stressful or sub-optimal environment. This issue is highly relevant to the welfare of the horse as equine stereotypies are often physically prevented through the application of invasive surgery, or through devices such as the crib-strap or anti-weave bar (McBride and Long, 2001). If these behaviours do serve as a coping response, then physical prevention may leave the animal at risk from the deleterious effects of stress (e.g. gastric ulceration, immunodeficiency).

The purpose of this paper, therefore, is to review current research in order to draw conclusions about the function of equine stereotypies. Work carried out in other species as well as the horse will be assessed and will be related to three areas of study, 1) the effect of stereotypy performance and restriction on stress-related parameters, 2) behavioural assessment of function and 3) the underlying neurochemical mechanism of stereotypy.

The effect of stereotypy performance and restriction on stress-related parameters

The most commonly employed approach to assess the function of stereotypy is to measure physiological stress related parameters, such as hypothalamo-pituitary-adrenal (HPA) or sympatho-adrenal (SA) activity, either pre- and post-stereotypy performance or pre- and post-prevention. Theoretically, if stereotypy has a coping function, then prevention will induce a rise in stress-related parameters, whereas stereotypy performance should result in a reduction. Research has been carried out in a number of species including the horse and has previously been summarised (McBride, 2000). It is presented again here in Table 6.1.

It is apparent that a certain amount of contradiction exists in the results for both equine and non-equine species. For example, HPA activity has been shown to both decrease (Dantzer *et al.*, 1988; Tazi *et al.*, 1986) and increase (Wallace *et al.*, 1983) as a result of stereotypy performance by rats. In the horse, HPA activity (cortisol) was significantly higher prior to crib-biting onset followed by a significant reduction (McBride and Cuddeford, 2001). Heart rate (an index of SA activation) has also been reported to reduce significantly following cribbing bouts (Lebelt, 1998). Conversely, prevention of crib-biting by removal of cribbing surfaces had no effect on HPA response (McGreevy and Nicol, 1995), whilst prevention of crib-biting and weaving by crib-strap and anti-weave grid respectively, elicited a significant increase in HPA activity in both stereotypy and control populations (McBride and Cuddeford, 2001). Thus, results from equine stereotypy performance studies seem to support the coping

Species and stereotypic behaviour	Effect of performance		Source
	HPA activity	SA activity	
Children, stereotypic leg-swinging		↓	(Soussignan and Koch, 1985)
Calves, tongue-play		↓	(Seo et al., 1998)
Pigs, oral stereotypies	↓		(Dantzer and Mormede, 1981)
Pigs, oral stereotypies		-	(Schouten et al., 1991)
Rats; schedule-induced polydipsia (SIP)	↓		(Brett and Levine, 1979)
Rats; SIP	↑		(Wallace et al., 1983)
Rats; SIP	↓		(Tazi et al., 1986)
Rats; SIP	↓		(Dantzer et al., 1988)
Horses; crib-biting		↓	(Lebelt, 1998)
Horses; crib-biting and weaving	↓		(McBride and Cuddeford, 2001)

Table 6.1 The physiological effects of the performance of stereotypic behaviour and its prevention (↓ decrease, ↑ increase, - no effect) on hypothalamo-pituitary-adrenal (HPA) and sympatho-adrenal (SA) activity.

Species and stereotypic behaviour	Method of stereotypy prevention	Effect of prevention		Source
		HPA activity	SA activity	
Rats; SIP	water removal	↑		(Tazi et al., 1986)
Rats; SIP	water removal	↑		(Dantzer et al., 1988)
Pigs; SIP	water removal	↑	-	(Dantzer et al., 1987)
Pigs; SIP	chain removal	↑		(Dantzer et al., 1987)
Pigs; drinking and post-prandial chain pulling	nipple drinker and chain removal	-		(Terlouw et al., 1991)
Pigs; post-prandial chain pulling	chain removal	-		(Schouten et al., 1991)
Bank-voles; stereotypic jumping		↑		(Kennes and De Rycke, 1988)
Mice; wire gnawing	modification of lid	↑acute (1 day); - chronic (10 days)		(Wurbel and Stauffacher, 1996)
Horses; crib-biting	removal of crib-biting surfaces	-		(McGreevy and Nicol, 1995)
Horses; crib-biting	Crib-strap	↑		(McBride and Cuddeford, 2001)

hypothesis, whilst the two prevention experiments are harder to interpret. The absence of a significant HPA response when cribbing surfaces were removed (McGreevy and Nicol, 1995) could have been affected by the stage of stereotypy development; the intensity of stereotypy performance (as a factor of development) appears to be important from a functional perspective. For example, in a 21-day trial conducted by Dantzer *et al.* (1987) using pigs performing schedule induced chain manipulation, those animals that were denied chain access on day 8 showed no change in HPA parameters compared to a chain provided group. However, by day 21 there was a trend towards higher plasma cortisol post-prevention. Between day 8 and 21, stereotypy score had risen from 800 to 1300 units. This suggests that a coping function is present only when the adjunctive behaviour becomes well established and is performed at relatively high intensities.

In conclusion, research measuring stress related parameters such as HPA or SA activity either pre- and post-stereotypy performance or pre- and post- prevention are yet to firmly conclude on the putative function of stereotypy, either in the horse, or other species. The significant HPA elevation seen in control and stereotypy groups where prevention was attempted via crib-strap and anti-weave bar indicates that these devices may be of concern from a welfare perspective.

The behavioural assessment of function

Using a two choice experiment Cooper and Nicol (1991) tested whether stereotypy changed the way in which an animal perceived its environment. Two populations of bank voles (stereotypy performing and non-stereotypy) were given the choice of an enriched versus barren environment. The non-stereotypy animals spent significantly more time in the enriched environment whilst the stereotypy group showed reduced preference. The authors concluded that stereotypy-performing animals did in fact perceive the barren environment as being less aversive than the non-stereotypy group and thus stereotypy was interpreted as having a coping function. This view has since been criticised, as the enriched environment may have impeded stereotypy performance (Rushen, 1993). However, the study still suggests that the animal is highly motivated to perform stereotypy and will suffer an aversive environment to facilitate its performance.

The second behavioural approach to elucidating stereotypy function relies on the assumption that if a commodity is rewarding, then individuals will perform work to access that commodity (Dawkins, 1991). Based on this, Falk (1971) reported that rats could be taught an operant task in order to engage in schedule-induced stereotypy. More recently, Smith *et al.* (2003) repeated this experiment in a preliminary study using a group of 7 crib-biting, and 7 non-stereotypy horses. In a stable devoid of cribbing surfaces all animals were shaped to press a button for a food reward. The experiment began when the food reward was replaced with the emergence of a cribbing surface following a button press. The cribbing surface was available for 15 seconds before retraction within the device. In a series of 5 8-minute trials conducted over 5 days (1 trial per day) crib-biting animals performed significantly more button presses than controls (P<0.01, t=5.54, DF 7) (see Figure 6.1).

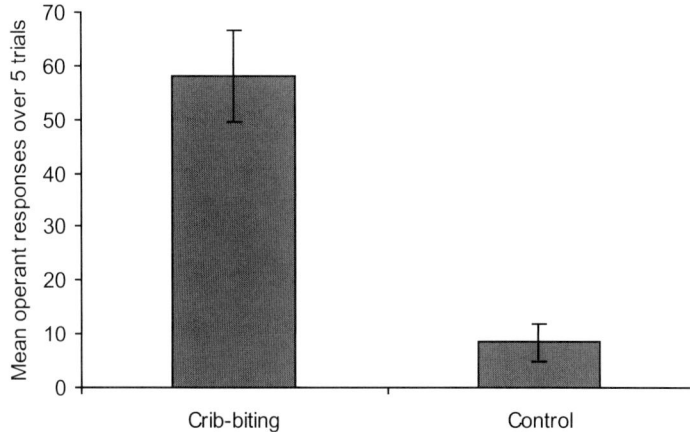

Figure 6.1.
Mean operant
responses over five
8 minute trials for
crib-biting and
control groups.

Crib-biting horses continued to respond over the 5-day period at a rate of at least 5 presses per 8-minute trial and directed clear cribbing behaviour at the bar when available. Total extinction of the operant response was seen after 2 trials in the control group. These results can be interpreted to suggest that horses will work in order to crib-bite, and that crib-biting, thus, has rewarding properties. However, an alternative interpretation of the results may exist based on results from spatial learning experiments conducted using bank voles (Garner and Mason, 2002). Bank voles that performed stereotypy at a high intensity were reported to have a significantly increased latency to extinction of an operant response compared to low intensity individuals. Garner and Mason (2002) concluded that lack of extinction could be related to a disinhibition of behavioural response selection mechanisms, occurring at the level of the striatum brain region. Potentially, this would cause the animal to continue a previously learned response despite loss of positive reinforcement. Thus, the continued button pressing in crib-biting horses may not be indicative of work performed in return for an opportunity to crib-bite. Instead, continued responding could simply indicate lack of extinction of the initial food motivated behaviour. Anecdotal evidence suggests the this same phenomenon indeed applies to the horse (see Nicol, 2002, for review).

In conclusion, current data from behavioural work on the function of stereotypic behaviour is open to several alternative interpretations.

The underlying neurochemical mechanism of stereotypy

Initial efforts to elucidate neurochemical pathways underlying stereotypy in farm animals concentrated on the administration of neurochemical antagonists. For example, Cronin et al. (1986) administered an opiate antagonist to pigs which resulted in a significant decline in stereotypic chain chewing. Opiate antagonists have also been reported to cause a complete cessation (Dodman et al., 1987) and an 84% reduction (McBride and Cuddeford, 2001) in crib-biting behaviour as well as a 30% reduction in equine weaving (Nurnberg et al., 1997), all of which indicate that opioid pathways are involved in stereotypy performance.

These results were interpreted by Cronin et al. (1986) to suggest that stereotypy is associated with central endogenous opioid release, and thus indicative of a coping function as opioids are known to posses hedonic properties. This was supported by a similar study by Schouten and Rushen (1992) who concluded that opiate antagonists inhibit an opioid mediated positive feedback pathway activated by stereotypy performance. However, the opioid antagonist theory of stereotypy function has been subjected to widespread criticism in the academic press, where various authors (e.g. Dantzer, 1991; Zanella et al., 1996) argued that if stereotypy has an opioid releasing effect, then blocking of opioid receptors within central nervous system reward centres, should initially lead to a higher stereotypy intensity as the animal attempts to maintain pre-administration levels of reinforcement. The general report of an immediate reduction post-administration thus supports the view that opioids are not so much the effect, but the cause of stereotypic behaviour. This may, however, still be an over-simplification of the underlying mechanism involved as administration of exogenous opioids are known to increase the motivation to perform rewarding behaviours, such as eating and sexual behaviours (Bakshi and Kelley, 1994; Band and Hull, 1990; Majeed et al., 1986; Mucha and Iversen, 1986; Wise, 1996), and similar behaviours can be attenuated by opiate antagonist administration (Giraudo et al., 1993). Given that these agonists and antagonists have the same effect on stereotypic behaviour (see McBride, 2000 for review), this would support stereotypy being classified as a reward behaviour with a potential functional role.

In an effort to further examine the link between opioid physiology and equine stereotypy, peripheral blood levels of beta endorphin have been compared during-cribbing and non-cribbing periods in crib-biting horses but with no significant difference reported (Gillham et *al.*, 1994; Lebelt, 1998; Pell and McGreevy, 1999). Other studies have reported stereotypy animals to have significantly higher (Lebelt, 1998), lower (Gillham *et al.*, 1994) and statistically similar (McBride and Cuddeford, 2001; Pell and McGreevy, 1999) basal levels of beta-endorphin to non-cribbing controls. Overall, this line of investigation has yielded inclusive results. Considering that most types of endogenous opioids are released, and have their receptors within the central nervous system (Smith and Lee, 2003), using peripheral measures as a reflection of central activity is unlikely to produce consistent data. Measures within the central nervous system, may be more informative in this respect.

Most complex behaviours are ordered by computational neural circuitry which receives direct input from internal and external sensory arrays. Deviation from homeostatic 'set points' leads to motivation towards a behaviour which re-establishes physical and mental equilibrium. Actions which bring an organism closer to equilibrium ('appetitive behaviours') are positively reinforced (rewarded) to ensure continuation until balance is achieved (Carlson, 2001). Brain centres responsible for behavioural motivation and reward encompass structures of the mesencephalon such as the ventral tegmental area (VTA), which projects dopamine containing nerve fibres (the A10 bundle) towards an area of the telencephalon known as the nucleus accumbens (Berridge, 1996). In fact, generation of reward sensation has been extensively correlated with activation of dopaminergic circuits in these brain regions. For example, microdialysis experiments have paired reinforcing events such as eating following food deprivation (Hernandez *et al.*, 1988) and sexual contact (Damsma *et al.*, 1990) with dopamine release in the VTA and accumbens. Moreover, a reward function is implied by conditioning studies which report that rats can be taught to press a lever that delivers minute injections of dopamine into the same regions (Hoebel *et al.*, 1983). In addition to dopaminergic innervation, the VTA also receives input from glutamatergic neurons of the pre-frontal cortex (PFC) (see Tzschentke, 2000, for review). Stimulation of the glutamatergic terminals from the PFC, causes activation of the VTA, and subsequent dopamine efflux into the accumbens (Gariano and Groves, 1988). In the context of behavioural selection, the PFC has been linked to the control of goal directed strategies (Krawczyk, 2002). In particular, appetitive behaviours which bring the animal closer to the consummatory goal, cause activation of the PFC (Matsumoto *et al.*, 2003). Subsequent activation of reward circuitry of the VTA and thus the nucleus accumbens, would then reinforce successful behaviour, thereby increasing the probability of goal achievement with minimal trial and error (Carlson, 2001). By examining the neurochemical interaction between highly motivated stereotypies and reward centres such as the nucleus accumbens, the function of these behaviours may be further elucidated. In the previous paper, stereotypies were defined as highly motivated, appetitive behaviours that are the culmination of a neural sensitisation process induced by stress or predisposed by genotype and are performed in an environment where consummatory aspects of the behaviours cannot be satisfied. This sensitisation process is considered to involve down regulation of inhibitory D2 auto-receptors on the VTA, and an up-regulation of excitatory D1 and D2 dopamine receptor subtypes on the nucleus accumbens compared to pre-stress values (Cabib *et al.*, 1998). From a functional perspective, sensitisation of accumbens dopamine transmission has been proven to lower reward threshold (Carr, 2002; Jones *et al.*, 1990) i.e. the same amount of stimulation may result in enhanced reward sensation. Thus, in the context of the sensitised phenotype, the performance of appetitive behaviours may have enhanced reward characteristics. Sufficient sensitisation may in fact give these behaviours reward qualities similar to that generated during the consummatory phase of goal-directed behaviours. This would potentially allow the animal to use the behaviours to:

1. self-stimulate in an environment lacking stimulation or, be used to counter specific stressors ('self-stimulation');
2. substitute other unobtainable consummatory behaviours ('substituted behavioural need').

Preliminary results on dopamine receptor physiology indicate that stereotypy horses have undergone a similar sensitisation process (McBride and Hemmings, 2001), suggesting that equine stereotypies may have similar enhanced reward characteristics and thus a putative function in this respect.

Conclusion

In conclusion, studies using inferred measures of stress to elucidate the functional role of stereotypies have, to date, generated contradictory results. Behavioural studies are also yet to fully prove or disprove any hypothesis in this respect. However, when placed in the context of other neurochemical research into the reward characteristics of appetitive behaviours and stress-induced reduction of reward threshold, the fact that stereotypy horses have altered dopamine physiology lends support to the idea that these equine stereotypies have reward characteristics and thus could be used by the animal from a functional perspective.

References

Bakshi, V. P. and Kelley, A. E. (1994). Sensitization and conditioning of feeding following multiple morphine microinjections into the nucleus-accumbens. *Brain Research* **648**: 342-346.

Band, L. C. and Hull, E. M. (1990). Morphine and dynorphin(1-13) microinjected into the medial preoptic area and nucleus accumbens - effects on sexual behavior in male rats. *Brain Research* **524**: 77-84.

Berridge, K. C. (1996). Food reward - brain substrates of wanting and liking. *Neuroscience and Biobehavioral Reviews* **20**: 1-25.

Brett, L. P. and Levine, S. (1979). Schedule-induced polydipsia suppresses pituitary-adrenal activity in rats. *Journal of Comparative Physiology and Psychology* **93**: 946-956.

Cabib, S., Giardino, L., Calza, L., Zanni, M., Mele, A. and Puglisiallegra, S. (1998). Stress promotes major changes in dopamine receptor densities within the mesoaccumbens and nigrostriatal systems. *Neuroscience* **84**: 193-200.

Carlson, N., R. (2001). *Physiology of behavior,* Allyn and Bacon, Boston, USA.

Carr, K. D. (2002). Augmentation of drug reward by chronic food restriction: Behavioral evidence and underlying mechanisms. *Physiology and Behavior* **76**: 353-364.

Commission of the European Communities (1983). Abnormal behaviour in farm animals.

Cooper, J. J. and Nicol, C. J. (1991). Stereotypic behavior affects environmental preference in bank voles, clethrionomys-glareolus. *Animal Behaviour* **41**: 971-977.

Cronin, G. M., Wiepkema, P. R. and Vanree, J. M. (1986). Endorphins implicated in stereotypies of tethered sows. *Experientia* **42**: 198-199.

Damsma, G., Pfaus, J. G., Nomikos, G. G., Phillips, A. G. and Fibiger, H. C. (1990). Sexual-behavior enhances dopamine transmission in the nucleus- accumbens and the striatum of the male-rat. *Psychopharmacology* **101**: S12-S12.

Dantzer, R. (1991). Stress, stereotypies and welfare. *Behavioural Processes* **25**: 95-102.

Dantzer, R. and Mormede, P. (1981). Pituitary-adrenal consequences of adjunctive activities in pigs. *Hormones and Behavior* **15**: 386-395.

Dantzer, R., Gonyou, H. W., Curtis, S. E. and Kelley, K. W. (1987). Changes in serum cortisol reveal functional differences in frustration-induced chain chewing in pigs. *Physiology and Behavior* **39**: 775-777.

Dantzer, R., Terlouw, E. M. C., Mormede, P. and Le Moal, M. (1988). Schedule-induced polydipsia experience decreases plasma corticostereone levels but increases plasma prolactin levels. *Physiology and Behavior* **43**: 275-279.

Dawkins, M. S. 1991. From an animals point-of view - motivation, fitness, and animal-welfare. *Behavioral and Brain Sciences* **14**: 753-753.

Dodman, N. H., Shuster, L., Court, M. H. and Dixon, R. (1987). Investigation into the use of narcotic-antagonists in the treatment of a stereotypic behavior pattern (crib-biting) in the horse. *American Journal of Veterinary Research* **48**: 311-319.

Falk, J. L. (1971). The nature and determinants of adjunctive behaviour. *Physiology of Behaviour* **6**: 577-588.

Gariano, R. F. and Groves, P. M. (1988). Electrical-stimulation of medial prefrontal and cingulate cortex elicits bursting in a small population of mesencephalic dopamine neurons. *Annals of the New York Academy of Sciences* **537**: 505-507.

Garner, J. P. and Mason, G. J. (2002). Evidence for a relationship between cage stereotypies and behavioural disinhibition in laboratory rodents. *Behavioural Brain Research* **136**: 83-92.

Gillham, S. B., Dodman, N. H., Shuster, L., Kream, R. and Rand, W. (1994). The effect of diet on cribbing behavior and plasma beta- endorphin in horses. *Applied Animal Behaviour Science* **41**: 147-153.

Giraudo, S. Q., Grace, M. K., Welch, C. C., Billington, C. J. and Levine, A. S. (1993). Naloxone's anorectic effect is dependent upon the relative palatability of food. *Pharmacology Biochemistry and Behavior* **46**: 917-921.

Hernandez, L., Lee, F. and Hoebel, B. G. (1988). Microdialysis in the nucleus accumbens during feeding or drugs of abuse - amphetamine, cocaine, and phencyclidine. *Annals of the New York Academy of Sciences* **537**: 508-511.

Hoebel, B. G., Monaco, A. P., Hernandez, L., Aulisi, E. F., Stanley, B. G. and Lenard, L. (1983). Self-injection of amphetamine directly into the brain. *Psychopharmacology* **81**: 158-163.

Hughes, B. O. and Duncan, I. J. H. (1988). The notion of ethological need, models of motivation and animal-welfare. *Animal Behaviour* **36**: 1696-1707.

Jones, G. H., Marsden, C. A. and Robbins, T. W. (1990). Increased sensitivity to amphetamine and reward-related stimuli following social-isolation in rats - possible disruption of dopamine-dependent mechanisms of the nucleus-accumbens. *Psychopharmacology* **102**: 364-372.

Kennes, D. and De Rycke, P. H. (1988). The influence of the performance of stereotypies on plasma corticosterone and eosinophil levels in bank voles (*Clethrionomys glareolus*). *Proccedings of the 29th International Congress of the International Society for Applied Ethology* 238-240.

Krawczyk, D. C. (2002). Contributions of the prefrontal cortex to the neural basis of human decision making. *Neuroscience and Biobehavioral Reviews* **26**: 631-664.

Lebelt, D. (1998). Physiological correlates associated with cribbing behaviour in horses: changes in thermal threshold, heart-rate, plasma beta-endorphin and serotonin. *Equine Veterinary Journal*

Supplement: 21-27.

Majeed, N. H., Przewlocka, B., Wedzony, K. and Przewlocki, R. (1986). Stimulation of food-intake following opioid microinjection into the nucleus accumbens septi in rats. *Peptides* 7: 711-716.

Mason, G. J. (1991). Stereotypies- a critical review. *Animal Behaviour* 41: 1015-1037.

Matsumoto, K., Suzuki, W. and Tanaka, K. (2003). Neuronal orrelates of goal-based motor selection in the prefrontal cortex. *Science* 301: 229-232.

McBride, S. D. (2000). An investigation into the stereotypic behaviour of the horse. *PhD thesis, University of Edinburgh.*

McBride, S. D. and Long, L. (2001). The perception and subsequent management of equine stereotypic behaviour by horse owners; implications for animal welfare. *The Veterinary Record* 148: 799-802.

McBride, S. D. and Cuddeford, D. (2001). The putative welfare-reducing effects of preventing equine stereotypic behaviour. *Animal Welfare* 10: 173-189.

McBride, S. D. and Hemmings, A. (2001). Striatum D1 dopamine receptors are significantly higher in horses performing stereotypic behaviour. *Research in Veterinary Science.* 70: 12.

McGreevy, P. D. and Nicol, C. J. (1995). Behavioural and physiological consequences associated with prevention of cribbing. *Proccedings of the 29th International Congress of the International Society for Applied Ethology* 135-136.

Mucha, R. F. and Iversen, S. D. (1986). Increased food-intake after opioid microinjections into nucleus accumbens and ventral tegmental area of rat. *Brain Research* 397: 214-224.

Nicol, C. J. (2002). Equine learning: progress and suggestions for future research. *Applied Animal Behaviour Science* 78: 193-208.

Nurnberg, H. G., Keith, S. J. and Paxton, D. M. (1997). Consideration of the relevance of ethological animal models for human repetitive behavioral spectrum disorders. *Biological Psychiatry* 41: 226-229.

Pell, S. M. and McGreevy, P. D. (1999). A study of cortisol and beta-endorphin levels in stereotypic and normal Thoroughbreds. *Applied Animal Behaviour Science* 64: 81-90.

Rushen, J. (1993). The coping hypothesis of stereotypic behavior. *Animal Behaviour* 45: 613-615.

Schouten, W. and Rushen, J. (1992). Effects of naloxone on stereotypic and normal behavior of tethered and loose-housed sows. *Applied Animal Behaviour Science* 33: 17-26.

Schouten, W., Rushen, J. and Depassille, A. M. B. (1991). Stereotypic behavior and heart-rate in pigs. *Physiology and Behavior* 50: 617-624.

Seo, T., Sato, S., Kosaka, K., Sakamoto, N. and Tokumoto, K. (1998). Tongue-playing and heart rate in calves. *Applied Animal Behaviour Science* **58**: 179-182.

Smith, A. P. and Lee, N. M. (2003). Opioid receptor interactions: Local and nonlocal, symmetric and asymmetric, physical and functional. *Life Sciences* **73**: 1873-1893.

Smith, N., Hemmings, A. and McBride, S. D. (2003). The potential reward value of equine stereotypic behaviour. *Equine Science Conference* Warwick, UK.

Soussignan, R. and Koch, P. (1985). Rhythmical stereotypies (leg-swinging) associated with reductions in heart rate in normal school children. *Biological Psychology* **21**: 161.

Tazi, A., Dantzer, R., Mormede, P. and Lemoal, M. (1986). Pituitary-adrenal correlates of schedule-induced-polydipsia and wheel running in rats. *Behavioural Brain Research* **19**: 249-256.

Terlouw, E. M. C., Lawrence, A. B., Ladewig, J., Depassille, A. M., Rushen, J. and Schouten, W. G. P. (1991). Relationship between plasma-cortisol and stereotypic activities in pigs. *Behavioural Processes* **25**: 133-153.

Tzschentke, T. M. (2000). The medial prefrontal cortex as a part of the brain reward system. *Amino Acids* **19**: 211-219.

Wallace, M., Singer, G., Finlay, J. and Gibson, S. (1983). The effect of 6-ohda lesions of the nucleus accumbens septum on schedule-induced drinking, wheelrunning and corticosterone levels in the rat. *Pharmacology Biochemistry and Behavior* **18**: 129-136.

Wise, R. A. (1996). Addictive drugs and brain-stimulation reward. *Annual Review of Neuroscience* **19**: 319-340.

Wurbel, H. and Stauffacher, M. (1996). Prevention of stereotypy in laboratory mice - effects on stress physiology and behavior. *Physiology and Behavior* **59**: 1163-1170.

Zanella, A. J., Broom, D. M., Hunter, J. C. and Mendl, M. T. (1996). Brain opioid receptors in relation to stereotypies, inactivity, and housing in sows. *Physiology and Behavior* **59**: 769-775.

7

Equine neutrophil elastase 2A concentrations in tracheal wash preparations and synovial fluid as a potential objective marker for inflammatory airway and joint diseases in horses

M P Dagleish
Moredun Research Institute, Pentlands Science Park, Bush Loan, Penicuik, Near Edinburgh EH26 0PZ, UK, Email: daglm@mri.sari.ac.uk

Introduction

Neutrophil elastase (NE), a powerful proteolytic enzyme found within the primary cytoplasmic granules of polymorphonuclear leukocytes (PMN), is capable of degrading almost any component of the extracellular pulmonary matrix and forms part of the cells' armoury against invading microbes (Crystal, 1996; Haslett, Savill and Meagher, 1989). It is essential during inflammation for diapedesis and migration to extravascular compartments (Henson and Johnson, 1987), oxygen independent bactericidal activity and the remodelling of damaged tissues (Ganz, Metcalf, Gallin and Lehrer, 1987) and is crucial for health as an absence results in increased susceptibility to pyogenic infections which, if untreated, are eventually fatal. Mature PMN contain a finite amount of NE (Campbell, Silverman and Campbell, 1989) which can only be synthesized during development within the bone marrow, especially the pro-myelocyte stage (Fouret, duBois, Bernaudin, Takahashi, Ferrans and Crystal, 1989).

PMN are often the first white blood cells to be attracted to an inflamed and/or infected area within the body and can persist there due to their role in remodelling damaged tissue. The presence of excess numbers of PMN is often used in horses as an indicator of bacterial infection in tracheal washes and synovial fluid or of active chronic obstructive pulmonary disease when in bronchoalveolar lavage fluid (BALF) or tracheal aspirates (Dixon, Railton and Mcgorum, 1995; Tremaine, 2000). At present differential cytology is utilised because microbiological culture of synovial fluid samples is often unrewarding, even when bacterial infections are known to be present. Cytology only denotes presence, not the state of activation of the PMN which may be vital in assessing the true significance of the cells' presence.

Equine PMN have been shown to contain two types of NE, equine NE 2A and 2B (ENE2A and ENE2B respectively), of which, ENE2A is three times more abundant and of similar elastinolytic activity as human NE. Previous work has developed a specific sandwich ELISA to enable determination of concentration of ENE2A released by activated PMN (Dagleish *et al.*, 1999). The aim of this preliminary work was to determine if concentrations of ENE2A in various clinically accessible body fluids could be correlated with known inflammatory respiratory and joint disorders with respect to developing a predictive diagnostic test and also the possibility of using ENE2A concentration as a means of assessing response to treatment and point of recovery on a cellular level.

Materials and methods

Tracheal washes were taken under sterile conditions via a flexible fibre-optic endoscope from 70 Thoroughbred horses in race-training (22 with overt clinical signs, 7 poor performers and 41 routinely screened) and examined by conventional methods (mucus score, cytology and microbiological culture) and for ENE2A concentration of cell free supernatants after centrifugation (1000 x g, 7 minutes) by a previously published specific sandwich ELISA to ENE2A (Dagleish *et al.*, 1999).

Synovial fluid samples (n = 49 samples from 25 horses) were excess clinical samples from various anatomical limb joints from cases referred for lameness examination or horses euthanased for non-lameness or systemic inflammatory disorders and examined by conventional methods (cytology and microbiological culture and identification) and cell free supernatants also used to determine the ENE2A concentration.

ENE2A concentrations were determined in duplicate for each sample. Doubling dilutions (1/2 – 1/3840 for tracheal washes and 1/2 - 1/8000 for synovial fluid supernatants) were made to ensure readings fell within the range of the internal standard curve present on each 96 well plate.

Results

Irrespective of the presence of suspected pathological bacteria, horses with overt clinical respiratory signs had significantly ($p < 0.001$) higher concentration of ENE2A (846 ± 1372 ng/ml) in tracheal washes compared to poor performers (30.4 ± 45.6 ng/ml) and those routinely screened (106.5 ± 371.3 ng/ml).

M.P. Dagleish

Conclusions

Determination of ENE2A concentration in tracheal washes and synovial fluid may provide a repeatable and objective evaluation of the presence and severity of respiratory tract and joint inflammation, especially in sub-clinical disease. This may also provide a means of objectively determining when training should resume after joint or respiratory disease, the response to that training and consequently improve the health, welfare and performance of horses.

Table 7.1
Mean ± SD and ranges of equine neutrophil elastase 2A (ENE2A) concentrations in synovial fluid samples from various cohorts of septic and non-septic joints. Note that despite the wide range of ENE2A concentrations in the non-septic and septic joint groups the ranges do not overlap and are significantly (p<0.001) different.

Joint and status	n	Mean ± SD (ng/ml)	Range (ng/ml)
Non-septic joints	40	78.1 ± 97.8	4.7 – 485.3
Septic joints	9	4142.0 ± 4135.0	741.0 – 12016.0

References

Campbell, E.J., Silverman, E.K. and Campbell, M.A. (1989). Elastase and cathepsin-g of human-monocytes - quantification of cellular content, release in response to stimuli, and heterogeneity in elastase-mediated proteolytic activity. *Journal of Immunology* 143: 2961-2968.

Crystal, R.G. (1996). *Alpha 1-Antitrypsin Deficiency*, Marcel Dekker, Inc. New York, Basel, Hong Kong.

Dagleish, M. P., Pemberton, A. D., Brazil, T. J., McAleese, S. M., Miller, H. R. P. and Scudamore, C. L. (1999). Kinetics of Equine Neutrophil Elastase Release and Superoxide Anion Generation Following Secretagogue Activation: A Potential Mechanism for Antiproteinase Inactivation. *Veterinary Immunology and Immunopathology* 72: 257-275.

Dixon, P.M., Railton, D.I. and Mcgorum, B.C. (1995). Equine pulmonary disease: a case controlled study of 300 referred cases. Part 3: Ancillary diagnostic findings. *Equine Veterinary Journal* 27: 428-435.

Fouret, P., duBois, R.M., Bernaudin, J.F., Takahashi, H., Ferrans, V.J. and Crystal, R.G. (1989). Expression of the neutrophil elastase gene during human-bone marrow cell-differentiation. *Journal of Experimental Medicine* 169: 833-845.

Ganz, T., Metcalf, J.A., Gallin, J.I. and Lehrer, R.I. (1987). 2 genetic-disorders that affect human-neutrophils (pmn) are associated with deficiencies of microbicidal and cyto-toxic granule proteins. *Clinical Research* 35: A 424.

Haslett, C., Savill, J.S. and Meagher, L. (1989). The neutrophil. *Current Opinion In Immunology* 2: 10-18.

Henson, P.M. and Johnston, R.B. (1987). Tissue-injury in inflammation - oxidants, proteinases, and cationic proteins. *Journal of Clinical Investigation* **79**: 669-674.

Tremaine, H. (2000). Infection of equine joints and tendon sheaths. *In Practice* **22**: 262-273.

8

Development and application of modern breeding technologies to the horse

W R Allen
University of Cambridge Department of Clinical Veterinary Medicine, Equine Fertility Unit, Mertoun Paddocks, Newmarket, Suffolk CB8 9BH, UK

The application of modern breeding technologies to the practical breeding of horses has accelerated worldwide over the past two decades in all breeds and types of horses except the Thoroughbred. Artificial insemination (AI) with fresh extended semen is used routinely on most stud farms with considerable improvements in efficiency, fertility, disease control and management costs. The ability of cooled semen to retain high fertility for 24-48 hours enables its widescale shipment within and between countries by modern transport methods. Even the use of deep frozen semen is increasing rapidly, in conjunction with a single fixed-time insemination in relation to gonadotrophin-induced ovulation. The wide scale use of the ultrasound scanner to monitor follicular development and ovulation has markedly improved diagnosis of the optimum insemination time, with consequent improvements in conception rates. However, repeated scanning examinations at intervals of only 4-6 hours in order to carry out insemination of frozen-thawed semen after, rather than before, ovulation is illogical and inefficient.

Embryo transfer too has seen a marked upsurge in recent years, especially in champion polo ponies in Argentina and Brazil and in competing Sporthorses and racing Quarterhorses in North America and parts of Europe. Embryo recovery rates are high (80-90%) when fertile, (generally maiden) mares which have been inseminated with fresh or cooled semen and in which ovulation has been diagnosed accurately by daily ultrasound examinations and/or serum progesterone measurements, are flushed on the 7th or 8th day after ovulation. Surgical embryo transfer rates, *via* mid-line or flank laparotomy, are equally high when transferring within a donor: recipient synchrony range of 0-to-3 days. Pregnancy rates from non-surgical transfer *via* the cervix tend to be lower (50-80%) and are highly influenced by the experience and sterile practices of the operator, the tortuosity of the recipient mare's cervix and health of her endometrium. Induction of multiple ovulations by treatment with exogenous gonadotrophins is possible but is not yet a practical proposition, and successful deep-freezing of equine embryos also remains problematical due to the relatively advanced stage of their development by the time they reach the uterus, and are recoverable, on day 7 after ovulation. Impermeability of the unique equine blastocyst capsule to glycerol or other cryoprotectants may also add to this problem.

The advent of a number of new breeding technologies in the last 5 years holds exciting promise for horse breeding. For example, the repeated recovery of multiple oocytes from valuable donor mares by transvaginal ultrasound-guided ovum pick-up (OPU) for fertilisation, either by transfer to the Fallopian tube of a mated or inseminated recipient mare (GIFT), or by intracytoplasmic injection of a single spermatozoon (ICSI) that may have been frozen-thawed or sex-sorted into X- or Y chromosome-bearing populations by passage through a high speed fluorescence-activated cell sorter (FACS), prior to use. And most recently, videoendoscopic or "blind" deep uterine insemination techniques to achieve reasonable conception rates of 50-70% per cycle by the deposition of ultra low doses of spermatozoa (e.g. 1-10 million) directly onto the utero-tubal junction (UTJ) at the tip of the uterine horn ipsilateral to the ovary containing the pre-ovulatory follicle; the technique has particular potential with both frozen-thawed semen of moderate quality and sex- sorted fresh or frozen semen.

Thus, many exciting modern advances are coming on line currently which have the potential to improve greatly the fecundity and general productivity of the relatively few high class competing fillies and proven broodmares that exist in the Sporthorse breeding industry. It only remains now for equine veterinary clinicians to pick up these "new batons" and for discerning and determined Sporthorse owners to run with and apply them to good effect.

9

Mapping the grey gene in Thoroughbred horses

M. M. Binns and J. E. Swinburne
The Animal Health Trust, Lanwades Park, Kentford, Newmarket, Suffolk, CB8 7UU, UK

Introduction

During the past eight years a comprehensive genetic linkage map for the horse has been generated in Newmarket, through a fruitful collaboration between the Genetics section at the AHT and Professor Twink Allen at the TBA Equine Fertility Unit (Swinburne et al., 2000). In order to assess whether the map was adequate to start mapping disease traits in which we are interested, it was decided to attempt to map a clear, easily scoreable trait such as grey coat colour in the Thoroughbred horse. Grey coat colour is inherited as an autosomal dominant trait. Thoroughbred horses are fairly inbred, and as such they represent one of the more difficult horse breeds for genetic mapping. The majority of grey Thoroughbreds trace their ancestry back to Allcock's Arabian who was born in 1700, and horsemen have long noted an association between grey coat colour and the incidence of melanomas. It is hoped that characterizing the mechanism of grey coat colour formation might also lead to an improved understanding of the molecular basis of melanoma formation.

Material and methods

Extensive pedigree analysis was undertaken to identify an active grey Thoroughbred stallion, whose grey sire was still alive. Paris House, and his sire Petong, were suitable and blood samples from both animals together with 42 offspring (19 grey and 23 non-grey) and 31 of the dams were obtained. DNA was extracted from all samples and then PCR amplified with 253 microsatellite markers spanning the equine genome. A substantial number of the markers were multiplexed such that markers could be analysed simultaneously. Primers were fluorescently labelled and PCR products analysed on an ABI 3700 automated sequencer. Pedigree information and genotypes was initially entered into a Cyrillic database and then exported and analysed using MLINK in order to locate markers linked to the grey locus.

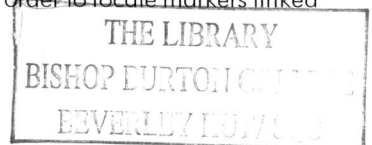

THE LIBRARY
BISHOP BURTON C
BEVERLEY

Results

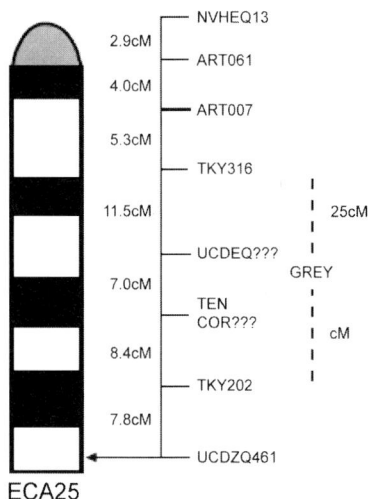

ECA25

Conclusions

The whole genome scanning set of horse microsatellite markers were effective in mapping an autosomal dominant trait, grey coat colour, even within one of the most in-bred horse breeds. This indicates that it should be possible to map more complex phenotypes in Thoroughbreds and other breeds.

References

Swinburne, J, Gerstenberg, C., Breen, M., Aldridge, V., Lockhart, L., Marti, E., Antczak, D., Eggleston-Stott, M., Bailey, E., Mickelson, J., Roed, K., Lindgren, G., von Haeringen, W., Guerin, G., Bjarnason, J., Allen, W.R. and Binns, M.M. (2000). First comprehensive low-density horse linkage map based on two, three-generation, full-sibling, cross-bred horse reference families. *Genomics*, **66**: 123-134.

Swinburne, J.E., Hopkins, A. and Binns, M.M. (2002). Assignment of the horse grey coat colour gene to ECA25 using whole genome scanning. *Animal Genetics*, **33**: 338-342.

10

Factors influencing placental and fetal development in the mare: important considerations for the selection of embryo recipients

S Wilsher and W R Allen
University of Cambridge, Department of Clinical Veterinary Medicine Equine Fertility Unit, Mertoun Paddocks Newmarket, Suffolk CB8 9BH, UK

In mammals the placenta is the vehicle for nutritional and gaseous exchange between the maternal and fetal circulations during pregnancy. The allantochorionic placenta of equids is classified as non-deciduate, non-invasive, epitheliochorial and diffuse and, as such, equine placentation demands fetomaternal interdigitation over the entire endometrial surface to optimise fetal development *in utero*. Any shortcomings are reflected by slower growth and delayed maturation of the fetus.

This requirement for an unabbreviated area of placental attachment is emphasised by the inability of mares to carry twin conceptuses to term due to simple competition between the two placentae for the limited area of endometrium which results in progressive starvation of one or both fetuses (Jeffcott and Whitwell, 1973). In addition, age-related chronic degenerative changes in the mare's endometrium (endometrosis) may compromise the uterine environment during pregnancy and Bracher *et al.* (1996) showed a close relationship between the health of the endometrium, the normality of structure and the density of the microcotyledons on the surface of the allantochorion, and fetal weight.

However, even in the case of a singleton pregnancy being carried by a mare with a healthy endometrium, other factors may influence placental development and, hence, fetal growth. For example, maternal size, and hence uterine size, profoundly effect the birthweight of the foal, as demonstrated by Walton and Hammond (1938) in their classical between-breed crossing of Shetland ponies and Shire horses, and by Tischner and Klimszak (1989) who transferred Pony embryos to large draft-type recipient mares and compared the birth size and subsequent development of the foals with sex-matched full siblings born from the genetic Pony mothers. In both experiments the size disparities between the foals at birth persisted into adulthood. Parity is also correlated with foal birthweight and Hintz *et al.* (1979)

noted that Thoroughbred mares < 7 years old produced foals with smaller cannon bones than mares aged 7-11 years old. Barron (1995) surveyed racetrack performance of Thoroughbreds and concluded that foals born from parous mares aged 7-11 years were more likely to be successful than those born from maiden mares or mares > 11 years. Similarly, Finocchio (1986) reported that third foals were most likely to become Stakes winners, followed closely by foals produced in parities 5,2 and 4, in that order.

In our laboratory, placentae from two cohorts of mares of varying size, age and parity were studied to elucidate some of the structural differences of the allantochorion that might influence foal birthweight. The first cohort utilised between- breed embryo transfer to create 8 Thoroughbred-in-Pony (Tb-in-P) pregnancies in which the genetically larger Thoroughbred fetus experienced cramping and nutritional deprivation *in utero*, 7 P-in- Tb pregnancies in which the smaller Pony fetus was exposed to nutritional excess *in utero*, with 7 normal Tb-in-Tb and 7 P-in-P pregnancies as controls. Strong correlations existed between maternal weight and foal birthweight and both the weight, volume and gross area of the allantochorion. The surface area of the microcotyledons, assessed stereologically, was influenced solely by the breed of the mother, with Thoroughbred mares having more plexiform microcotyledons than Pony mares. Compromised intrauterine conditions, as in the Tb-in-P pregnancies, led to lengthening of the microcotyledonary villi, with no increase in the degree of branching. The total microscopic area of fetomaternal contact across the entire placental interface was also calculated and showed a strong correlation with foal birthweight. These findings demonstrated convincingly that the *in utero* growth of the foal is governed by 'actual' placental size and competence which, in turn, is governed by the available area of healthy endometrium.

In the second experiment 84 placentae from commercial Thoroughbred mares were separated into 4 groups on the basis of age and parity; primiparous mares aged 4-7 years and multiparous mares aged 5-9 years, 10-15 years and ~ 16 years. Gross placental parameters were lowest in the maiden primiparous mares and of particular interest was the finding that microcotyledon surface area was reduced in these animals compared to the 5-9 y.o. multiparous mares. As expected, development of the microcotyledons was also constrained in the aged mares (>16 years old), presumably as a result of endometrosis in the apposing endometrium. In these older mares deficiencies in the microcotyledons was offset, to some degree, by a higher volume and gross area of the placenta which together maintained the total microscopic area of fetomaternal contact and, hence, foal birthweight. Thus, primiparity appears to be more significant in determining foal birthweight than maternal age, due to reductions in microcotyledon structure, coupled with reductions in the other gross placental parameters, together limiting the total area available for haemotrophic exchange across the placental interface.

Accordingly, when selecting recipient mares for use in a embryo transfer programme, it should be remembered that maternal parameters such as body size, age and parity will all exert varying degrees of influence on the health and total microscopic area of the allantochorion which, in turn, will govern the size and vitality of the foal at birth, and in later life. Ideally, recipient mares should be aged between 5 and 9, should have already produced at least one foal, and should have a body size or "roominess" that is equal to or greater than that of the donor mare.

References

Bracher, V., Mathias, S and Allen, W.R. (1996). Influence of chronic degenerative endometritis (endometrosis) on placental development in the mare. *Equine Veterinary Journal* **28**: 180-188.

Finocchio, E.J. (1986). Race performance and its relationship to birthrank and maternal age. *Proceedings of the American Association of Equine Practitioners* **31**: 571-578.

Jeffcott, L.B. and Whitwell, K.E. (1973). Twinning as a cause of foetal and neonatal loss in the Thoroughbred mare. *Journal of Comparative Pathology* **83**: 91-105.

Hintz, H.F ., Hintz, R.L. and Van Vleck, L.D. (1979). Growth rate of thoroughbreds. Effect of age of dam, year and month of birth, and sex of foal. *Journal of Animal Science* **48**: 480-487.

Tischner, M. and Klimszak, M. (1989). Development of Polish pony foals born after transfer to large mares. *Journal of Reproduction and Fertility* Supplement **8**: 62-63.

Walton, A and Hammond, J. (1938). The maternal effects on growth and conformation in Shire horse-Shetland crosses. *Proceedings of the Royal Society, Series B.125*: 311-335.

11

Towards a common equine rationing system across the European Union

W Martin-Rosset[1]
[1]*National Institute for Agricultural Research (INRA), Research Centre of Clermont-Ferrand, Theix, 63122 Saint-Genes-Champanelle, France*

Summary

Since the 1980's, two feed evaluation systems, the NRC and the INRA, have been proposed, suggesting allowances for energy and protein (Cuddeford, 1998). NRC recommendations include suggestions for digestible energy (DE) and digestible crude protein (DCP in 1978) and later crude protein (CP in 1989). INRA, (from 1984-1990), have firstly suggested the Net Energy System (NE), also referred to as the UFC system in French, (so called, as the NE system relies on expressing the NE value of a feed relative to a reference feed [barley], or Horse Feed Unit). The INRA have also gone on to recommend the Horse Digestible Crude Protein system (HDCP, or MADC in French). Whilst the INRA and NRC systems have both proposed nutritional requirements, the NRC is exclusive in its use of a protocol called the 'factorial method'. The factorial method was initially contrived to allow for the determination of physiological expenditure, where the efficiencies of the nutrients were known and relevant. Other influences could then be considered by implementing feed trials to determine allowances which would take into account most environmental factors. Both the NRC and the INRA have proposed a range of Dry Matter Intake (DMI) values, which intend to meet the nutritional requirements of differing types of horses. DMI values provided by the INRA have been drawn directly from feeding experiments conducted over the last 30 years. Ingestibility of the main types of forages have also been extensively measured in horses, using feeding trials devoted to that purpose. The following chapter will review existing methods and suggest a progression strategy towards a common rationing system.

Suggestions for a common format

The NRC and INRA systems have been implemented worldwide, but have been more commonly used in Europe where local adaptations are often made.

Within Europe, several countries use the Metabolisable Energy (ME) system, for example, in Finland, where energy is expressed in Feed Units, and in Sweden, where energy is expressed in joules. In Russia the Oat Feed Unit is used as an expression of energy (Memedeikin, 1990); whereas in Poland and Romania, the INRA systems are generally used. When considering protein recommendations, the DCP system is extensively used throughout the Northern European countries (Autsbo, 2002).

A comparison of NRC and INRA systems indicates that DE and DCP systems over-estimate energy and protein values by 5 – 35% and 5 – 20% respectively, according to the types of feeds used. DMI and the proportion of concentrate feed in the diet required, as stated by NRC, are, on average, significantly higher than INRA recommendations, irrespective of the type of horse. NRC requirements for energy and nitrogen are 2 – 17% and 2 – 36% lower respectively than those performed by INRA. These discrepancies are confirmed when feed allowances are matched between NRC vs INRA systems for the same diet (same forage, same concentrate, supplied in the same forage to concentrate ratio) offered to the same type of horses (Martin-Rosset and Tisserand, 2002; Martin-Rosset and Vermorel, 2002).

In light of the many differing systems available, it may be prudent to ask what could be suggested to enable a European rationing system to be implemented as standard across the EU?

The key process in achieving this goal would be to develop European working group. The scope of which could include:

• Sharing nutritional concepts using the same frameworks: GE, DE, ME, NE for energy: CP, DCP, HDCP for nitrogen; each country being free to implement through this common step-wise approach, the system they wish to use, up to the step they wish to reach according to local condition.
• To determine the more appropriate criteria, laboratory method and finally, tools (tables and equations) for evaluating, then predicting feed nutritive value, at each step of the energy and nitrogen frameworks.
• Comparison of husbandry systems, breed types, and existing rationing models.

Moreover, criteria, lab methods and tools should be referred to, and/or extended to include a wider range of feedstuffs, and food processing methods, which are implemented by the European feed industry. This could be carried out by using the data, and/or, sample banks available in the different countries or via new digestion trials, ring tests and calculations conducted within the scope of a European program designed by a working group. By way of an example, the digestible energy (DE) of forages could be predicted either by the use of an equation which relates DE to chemical composition (Zeyner and Kienzle, 2001), or by using an *in vitro* batch culture technique (Lowman *et al.* 1999), or finally by the use of an equation which would relate the digestibility of Energy (dE) to the digestibility of Organic Matter (dOM), as performed by INRA and PR, (Prakliklijkonderzoek Rundvee) (Martin-Rosset *et al.* 1994). In the latter study, dOM was predicted from chemical composition. Other methods include the pepsin cellulase method, or NIRS (Martin-Rosset *et al.* 1996). It should be relevant to compare the robustness and feasibility of these techniques by laboratories that implement them daily. In addition, it would be relevant to study the digestibility of silages, namely haylage, with 50-70% DM, as they are increasingly used for feeding horses in Northern Europe (Hale and Moore-Colyer, 2001), as well as temperate zones (Germany: Coenen *et al.* 2003; and France: Micol and Martin-Rosset, 1995). The exact equation to be used for accurately predicting energy value of 'alternative forages' should be discussed and determined by a group of experts.

The question also arises for Mediterranean forages as pointed out by preliminary work conducted in the Southern part of Italy by Miraglia *et al.*(2000) and Corino *et al.* (1996) . From this, it would then be relevant to investigate further the digestibility of concentrate feeds: ingredients and compound feeds, relative to feed processing techniques. Some such works are in progress in Italy, Finland, France, and in Sweden. The idea would be to construct European feed tables and sets of equations.

Similarly, requirements have to be refined using physiological data available in the literature. Firstly, it would be prudent to establish the parameters within which each country operates, allowing for differing requirements, allowances, and feed availability. Then the recommended allowances, which are the ones useful and relevant to the user, must be adjusted to local conditions, using the same nutritional models. Common feeding methods should be implemented to extend the availability of data banks, using new trials conducted under the auspices of a European working group. Using growing horses as an example, there are several nutritional models running in Europe, but it could be postulated that these models would not be transferable between countries, as the differences between the European horse breeds are too diverse. This said, it could be easy to compare the growth curve of the main European breeds for determining the potential discrepancy in maturity and/or precocity, going on to then to test the robustness of the different models.

The latter suggestions will hopefully spark useful discussions focussing on the design and integration of a common European rationing system. The work could be carried forward under the umbrella of the EAAP and the Horse Commission. It is envisaged that key members from the latter organisations will form a scientific committee to guide the process of change.

References

Autsbo, A. (2002). *The Scandinavian adaptation of the French UFC system*. In: Proceedings of the 1st European workshop on Equine Nutrition, Dijon, France 17-18 January, 69-78.

Cuddeford, D. (1998). *Feeding systems for Horses*. In: *Feeding systems and feed evaluation models*. CAB International, Wallingford, UK, pp 239-274.

Coenen, M., Mülller, G., Enbergs, H. (2003). Grass silages vs hays in feeding horses. In: Proceedings of the 18th ENPS symposium, Lansing, Michigan, USA, 4-7 June, 138-139.

Corino ,C., Miraglia, N., Polidori, M. (1996) Determination of the organic matter digestibility and prediction of the UFC value of Mediterranean forages. In: Proceedings of the 47th EAAP meeting Lillehammer, Norway, 25-29 August, Horse commission, Session 4, Abstract H4.3, p.292.

INRA, (1984). *Le Cheval* , Edited by R.Jarrige and W.Martin-Rosset, INRA Editions , 78026 Versailles, France, pp 687.

INRA, (1990). *Alimentation des chevaux*, Edited by W.Martin-Rosset, INRA Editions, 78026 Versailles, France, pp 232.

Hale, C,E., Moore-Colyer, M,J,S. (2001). Volountary food intake and apparent digestibilty of hay, big bales, silages and red clover by ponies. In: Proceedings of the 17th ENPS symposium, Lexington, USA, May 31-June 2,468-469.

Lowman, R.S., Hyslop, J.J., Longland, A.C., Cuddeford, D., Hollands, T. (1999). Evaluation of an in vitro batch culture technique for estimating the in vivo digestibility and digestible energy content of equine feeds using equine faeces as the source of microbial inoculum. *Animal Feed Science and Technology*, 85:11-27.

Martin-Rosset ,W., Vermorel, M., Doreau, M., Tisserand, J.L., Andrieu, A. (1994). The French horse feed evaluation systems and recommended allowances for energy and protein. *Livestock Production Science*, 40: 37-56.

Martin-Rosset, W., Andrieu, A., Vermorel, M. (1996). Routine methods for predicting the net energy value (UFC) of feeds in horses. In: Proceedings of the 47th EAAP meeting, Lillehammer, Norway,

22-29 August, Horse Commission, Session H4, Abstract H4.1p.292. Wageningen Pers, Wageningen, Netherlands.

Martin-Rosset, W., Vermorel, M. (2002). *Evaluation and expression of energy allowances and energy value of feeds in the UFC system for the performance horse*. In: Proceedings of the 1st European workshop on Equine Nutrition, Dijon, France, 17-18 January, 19-58.

Martin-Rosset, W., Tisserand, J,L. (2002). *Evaluation and expression of protein allowances and protein value of feeds in the MADC system for the performance horse*. In: Proceedings of the 1st European workshop on Equine Nutrition, Dijon, France, 17-18 January, 109-148.

Memedeikin, V.G. (1990). The energy and nitrogen systems used in USSR for horses. Proceedings of 41st EAAP meeting Toulouse, 9-12 July, France, Horse Commission, Session 3, Abstract H1.3p382.

Micol, D., Martin-Rosset, W. (1995). *Feeding systems for horses on high forage diets in the temperate zone*. In: Proceedings of the 4th International symposium on the Nutrition of Herbivores, Clermont-Ferrand, France, September 11-15, INRA Editions 78026 Versailles, France, 569-584.

Miraglia,N., Bergero, B., Gagliardi, D. (2000). *Il cavallo atleta*. Calderini edagricole edizioni, Bologna, Italy.

NRC, (1978). *Nutrients Requirements of Horses*. 5th Edition, National Academy Press , Washington D.C,USA, pp 33.

NRC, (1989). *Nutrients Requirements of Horses*. 6th Edition , National Academy Press, Washington D.C,USA, pp 100.

Zeyner, A., Kienzle, E. (2001). Ein neus Konzept zur energetischen Futterverwetung beim Pferd. Proc. Soc. Nutr. Phys., 10:106.

12

Partitioning degradation of feeds between different segments of the equine digestive tract

J. J. Hyslop
Dept Vet Clinical Studies, Edinburgh University, Easter Bush Vet Centre, Roslin, Midlothian EH25 9RG UK
Current address: ADAS Redesdale, Rochester, Otterburn, Newcastle upon Tyne, NE19 1SB, UK, *jimmy.hyslop@adas.co.uk*

Abstract

This paper describes how experimental adaptation of the mobile bag and in situ porous bag techniques for use in nutrition studies with equines, combined with mathematical modelling of feed degradation profiles and digesta passage rates, allows the quantitative partition of feed degradation amongst the major segments of the equine digestive tract. Dry matter effective degradability values in the small intestine (pre-caecal), caecal and colonic segments of the equine digestive tract for four fibre feeds are presented. Measurement of digesta passage rates in the caecum and total tract of equines are also discussed. Further studies are now required to capitalise on this development and partition degradation of individual feed constituents in common equine feeds across each of the major segments of the equine digestive tract. This would allow optimum, balanced and healthy diets to be formulated in practice across a wide range of equine husbandry systems.

Introduction and background

The equine digestive tract is physiologically arranged into the stomach, small intestine, caecum and colon. Although a limited degree of microbial fermentation occurs in the equine stomach, digestion of feed constituents pre-caecally is principally achieved by secretion of gastric acids and host animal enzymes into the lumen of the gut in a manner similar to monogastric species such as the pig. In contrast however, digestion of feed constituents within the hindgut (caecum and colon) of equines is achieved principally *via* microbial fermentation similar to that occurring in the rumen of cattle and sheep. In recent years, research carried out collaboratively by Edinburgh University, the Institute of Rural Studies (IRS) and the Institute of Grassland and Environmental Research (IGER) has studied these equine feed degradation processes in detail.

One vital focus of this research has been the adaptation of ruminant and pig nutritional research techniques for use within equine nutrition studies. Adaptation of these techniques allows both the dynamics and extent of feed degradation to be quantified within each of the major physiological segments of the equine digestive tract. Accurate estimation of feed degradation patterns over time are important since, along with digesta passage rate, it is both the extent and rate of feed degradation that determines how much of a particular nutrient that is digested in and absorbed from, each particular segment of the equine digestive tract. Knowledge of the amount of a particular nutrient that is available to the horse within each segment of the digestive tract from each feedstuff is critical in ensuring that equine rations are properly balanced. Well-formulated and balanced equine rations will optimise work outputs and feed efficiency as well as minimising digestive upsets and associated health problems.

The objective of this paper is to outline the experimental approaches that have been developed and to illustrate the results obtained from combined use of the mobile bag technique (MBT) and the *in situ* technique (IST) in equines.

Methodology

Techniques adapted from ruminant research for use in equines

The techniques adapted from ruminant nutrition research for use in equines during these studies are the mobile bag technique (MBT) and the *in situ* technique (IST). The principle use of the MBT, which led to its development, was to study the digestibility of protein in pig diets (Sauer *et al.*, 1983) and to study the post-ruminal digestibility of undegraded dietary protein in ruminant diets (Hvelplund, 1985). Similarly the IST has been used principally to study the ruminal digestion of feedstuffs (Mehrez and Ørskov 1977). In our work the MBT has been used to study feed degradation profiles in both the small intestine and total tract of equines whilst the IST has been used to obtain feed degradation profiles following incubation in the equine caecum.

Use of the mobile bag technique (MBT) in the equine small intestine

Full details of the methods used in our application of the MBT in equines including pre- and post-incubation sample processing are given in Moore-Colyer *et al.* (2002) and McLean *et al.* (1999). Similar procedures have been described by Macheboeuf *et al.* (1996).

Key aspects of MBT use in the small intestine of equines are as follows: mobile bags (6x1 cm Ø or 4x1 cm Ø) containing two small steel washers and the feed sample are placed in the stomach using a naso-gastric tube. They are subsequently recovered on a specially made magnetic device that inserts inside a caecal fistula positioned at the ileo-caecal junction (Cottrell *et al.*, 1998). Mobile bags are recovered over a period of time ranging from 1 - 8 hours after placement in the stomach, in effect yielding a range of incubation times within the pre-caecal segment of the equine digestive tract. Feed degradation profiles can then be fitted to losses from these mobile bags using existing mathematical degradation models (Ørskov and McDonald 1979, Dhanoa 1988). A typical dry matter (DM) degradation profile for unmolassed sugar beet pulp (USBP) in the small intestine of equines is given in Figure 12.1.

Small intestine USBP DM disappearance from mobile bags

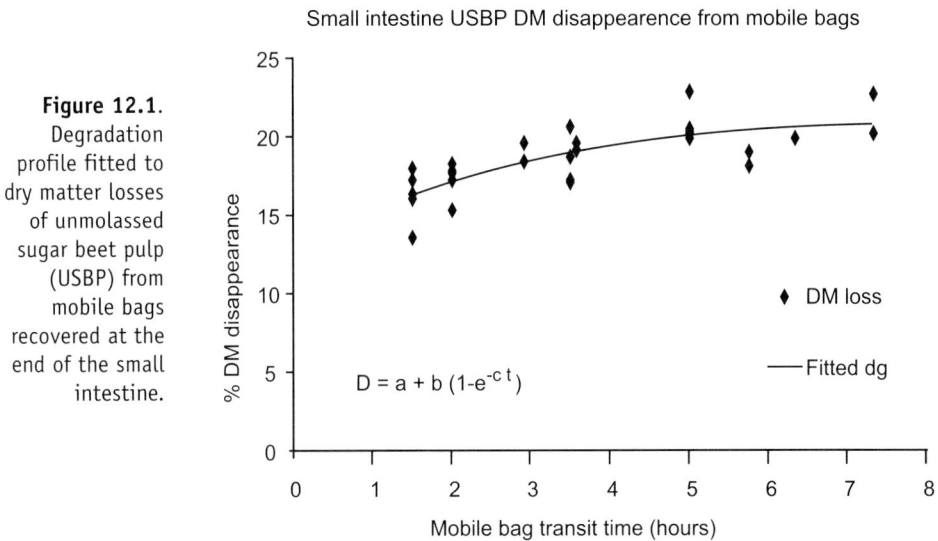

Figure 12.1. Degradation profile fitted to dry matter losses of unmolassed sugar beet pulp (USBP) from mobile bags recovered at the end of the small intestine.

$$D = a + b (1-e^{-ct})$$

Use of the *in situ* technique (IST) in the equine caecum

Full details of the methods used in our application of the IST in equines including pre- and post-incubation sample processing are given in Hyslop *et al.* (1999b). Similar procedures have been developed by others (Applegate and Herschberger, 1969; Drogoul *et al.*, 1995; Miraglia *et al.*, 1988).

Key aspects of the use of the IST in the caecum of equines are as follows: 6.5x20 cm *in situ* bags containing the feed samples are suspended in the caecum for a range of incubation times (e.g. 0,3,5,8,16,24,48 hours). We recommend the use of more than one *in situ* incubation sequence since results have shown that degradation profiles are sensitive to incubation sequence, probably due to variation within the caecal digesta pool (Hyslop *et al.*, 1999b). Existing feed degradation models (Ørskov and McDonald 1979, Dhanoa 1988) are again used to describe feed degradation profiles. Typical caecal *in situ* degradation profiles for four feedstuffs are shown in Figure 12.2.

Figure 12.2.
In situ degradation profiles of unmolassed sugar beet pulp (USBP), hay cubes (HC), soya hulls (SH) and a 2:1 mix of oat hulls:naked oats (OHNO) following incubation in the equine caecum.

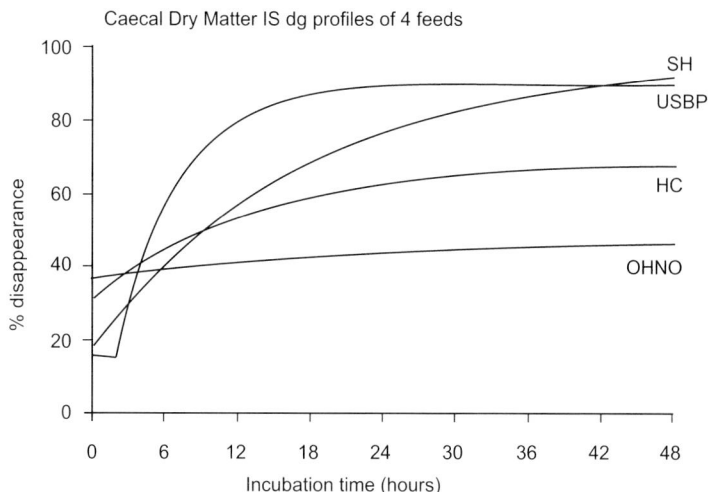

Caecal Dry Matter IS dg profiles of 4 feeds

Use of the mobile bag technique (MBT) in the equine total tract / colon

Full details of the methods used in our application of the MBT in the total tract of equines as a means of determining feed degradation profiles in the colon are given in Hyslop *et al.* (1998) and Moore-Colyer *et al.* (2002).

Key aspects of the use of the MBT in the total tract of equines are as follows: mobile bags (6x1 cm Ø or 4x1 cm Ø) without steel washers and containing the feed sample are placed in the stomach using a naso-gastric tube. They are subsequently allowed to travel through the whole digestive tract and are recovered in the faeces over a period of time, in effect yielding a range of incubation times within the whole digestive tract of equines. Feed degradation profiles can then be fitted to losses from these mobile bags using existing feed degradation models (Ørskov and McDonald 1979, Dhanoa 1988). A typical dry matter (DM) degradation profile for unmolassed sugar beet pulp (USBP) in the whole digestive tract of equines is given in Figure 12.3.

Combined degradation profiles in different segments of the equine digestive tract

These degradation profiles obtained from the MBT in the small intestine, the IST in the caecum and the MBT over the total tract of equines can now be combined to give a description of feed degradation in equines. Typical examples of combined degradation profiles for USBP and a ground and pelleted hay cube (HC) are given in Figures 12.4 and 12.5 respectively.

USBP DM disappearance from total tract mobile bags

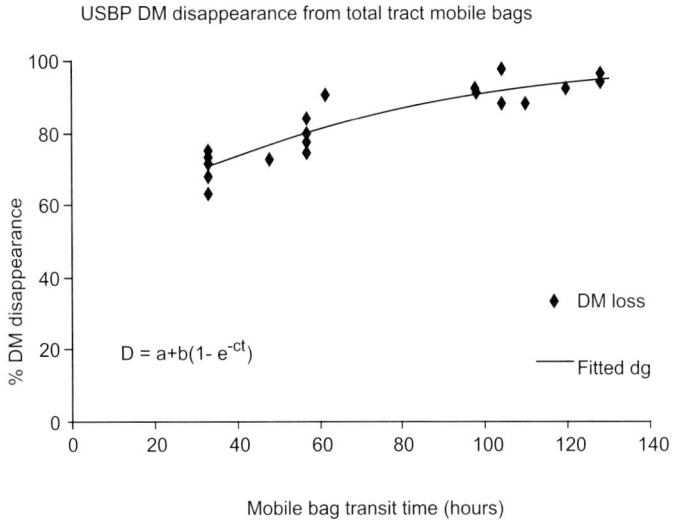

Figure 12.3. Degradation profile fitted to dry matter losses of unmolassed sugar beet pulp (USBP) from mobile bags recovered from faeces.

$D = a+b(1- e^{-ct})$

Mobile bag transit time (hours)

USBP DM dg *in situ* (caecum),small intestine and total tract MBT

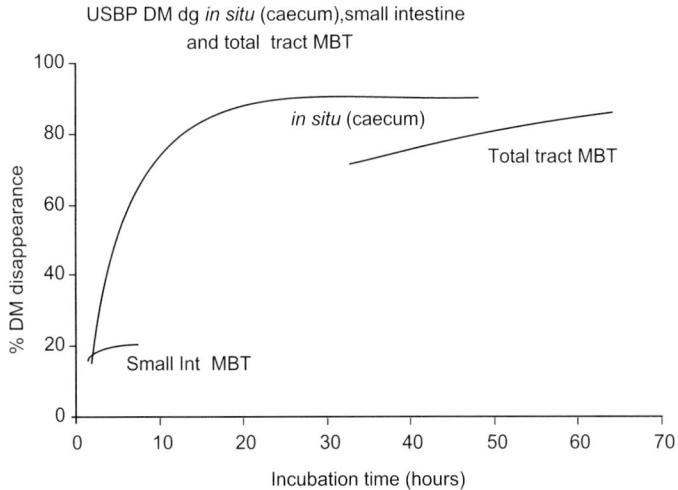

Figure 12.4. Combined unmolassed sugar beet pulp (USBP) dry matter degradation profiles.

Incubation time (hours)

Digesta passage through the equine digestive tract

It should be remembered that feed degradation profiles simply represent degraded feed particles that are lost from a porous bag during incubation within any given segment of the equine digestive tract. They do not represent any subsequent absorption, utilisation or passage of the degraded feed particles. Degraded feed particles that leave a porous bag can either be absorbed across the gut wall into the bloodstream of the animal (small intestine) or may pass out into

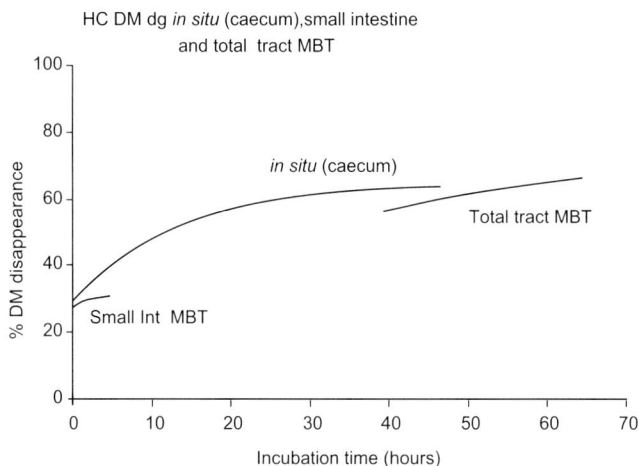

Figure 12.5.
Combined hay
cubes (HC) dry
matter degradation
profiles.

the next segment of the digestive tract *via* the normal processes of digesta passage as ingested feed is propelled through the gut. In the caecum and colon, feed particles or their degradation products may also be utilised (further digested) by microbes as a source of nutrients, a process which may itself produce further energy yielding substrates for the horse (i.e. volatile fatty acids). Digesta passage is a highly complex process but can also be modelled mathematically assuming either time-independent (Grovum and Williams 1973; Dhanoa *et al.*, 1985) or time-dependent models (Pond, *et al.*, 1988; Ellis *et al.*, 1994). Space does not allow a detailed discussion of digesta passage through the horse but some brief features are described below.

Digesta passage through the stomach and small intestine of equines

Digestion of feed (either in or out of a mobile bag) in the stomach and small intestine occurs while that feed (mobile bag) is passing through the segment. Feed that is degraded during that process of passage can either be absorbed across the gut wall into the bloodstream or can pass out into the caecum as it reaches the end of the small intestine. Feed that is degraded early in that process of passage will have a greater probability of absorption than feed which is degraded later during that process of passage since the feed degraded later will already have passed through much of the site of absorption (small intestine). The converse of this is that feed degraded later during this process of passage will have a higher probability of passage into the caecum than feed degraded during the early stages: i.e. the probability of passage is dependent on the time that feed is degraded during its passage through the small intestine. Figure 12.6 attempts to visually describe this process.

We do not have any experimental data available to establish conclusively the nature of digesta passage through the small intestine of equines. Clearly more research is needed in this area. However, the concept of time dependent digesta passage through tubular segments of the digestive tract in mammals (e.g. the small intestine) is not new (Ellis et al., 1994) and we would advocate its use in this context.

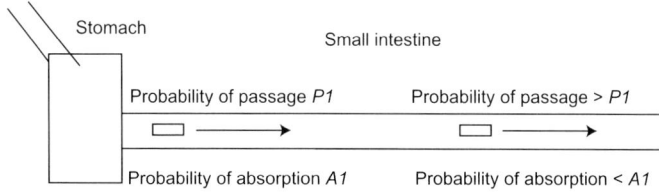

Figure 12.6. Digesta passage through the small intestine of equids.

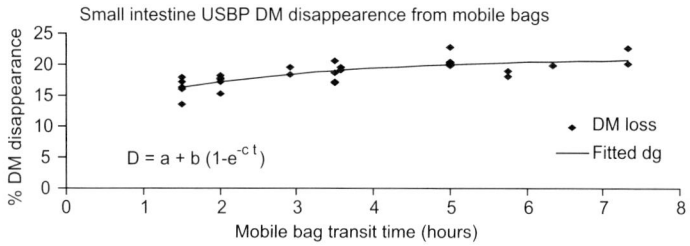

The equation shown in the figure: $D = a + b (1 - e^{-ct})$

Digesta passage through the caecum of equines

We have conducted studies on the passage of marked feedstuffs from the caecum of ponies (Hyslop et al., 1999a; McLean, 2001) and observed that marker concentration declines in a time-independent, exponential form similar to observations in the rumen of cattle and sheep. Figure 12.7 depicts the marker decline following a pulse dose of Chromium labelled Hay Cubes (HC) or Rolled Barley (RB) into the caecum of fistulated ponies. Given the accurate description of these marker declines with simple exponential relationships ($R^2 > 90\%$), it seems reasonable to proceed on the basis that the digesta passage through the caecum of equines is a time-independent process and adopt the appropriate mathematical modelling approaches in partitioning calculations.

Figure 12.7. Time-independent, exponential digesta passage from the caecum of equines following a pulse dose of Chromium labelled hay cubes or rolled barley.

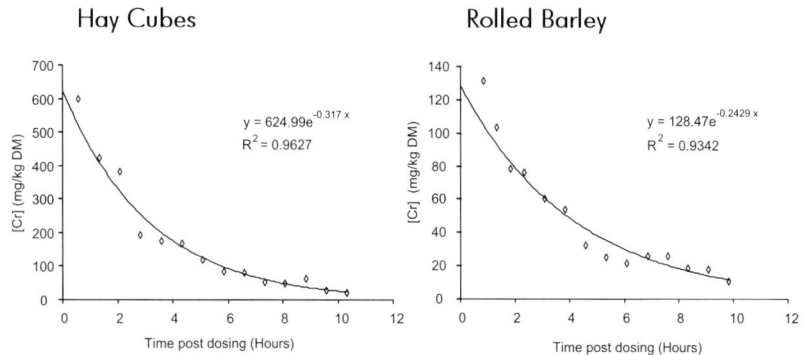

Hay Cubes: $y = 624.99e^{-0.317x}$, $R^2 = 0.9627$

Rolled Barley: $y = 128.47e^{-0.2429x}$, $R^2 = 0.9342$

Digesta passage through the total tract / colon of equines

We have also obtained some marker excretion curves for Ytterbium (Yb) and Dysprosium (Dy) labelled feeds following an orally administered pulse dose (Moore-Colyer, 2000; Hyslop, unpublished). Given these faeces marker concentration curves the applicability of several digesta passage models (Dhanoa, et al 1985, Grovum and Williams 1973, Pond et al 1988) can be examined. In general we have found that whilst the time-independent models of Dhanoa, et al. (1985) and Grovum and Williams, (1973) can be used to fit marker faecal excretion data, it is the time-dependent 2-compartmental models of Pond et al. (1988) that are the most successful in describing digesta passage through the total tract of equines. Similar observations and conclusions have recently been published by Moore-Colyer et al. (2003). Whilst definitive conclusions are yet to be drawn, these observations lend support to the view that digesta passage is likely to be a time-dependent process in at least some part of the equine digestive tract. Clearly more research and data analysis are needed in this area. Figure 12.8 shows an example of one such curve for a fibre-based diet (Moore-Colyer, 2000) with the two time-independent models (Dhanoa, et al., 1985; Grovum and Williams, 1973) fitted. Similarly, Figure 12.9 shows another example where the time-dependent 2-compartmental Gamma 3 (G3G1) model of Pond et al. (1988) has been fitted to the faecal excretion data of a Dy marked concentrate feedstuff (Hyslop, unpublished). More research and data analysis is needed before a conclusion regarding time dependent vs time independent passage through the equine colon / total tract can be arrived at.

Figure 12.10 depicts a simple attempt at providing a biological interpretation of what these marker excretion curves might mean. The "transit time" (time lag before first appearance of the marker in faeces) might reflect digesta passage through some tubular segments of the equine digestive tract. This may reflect transit through the stomach and small intestine of equines or alternatively it may reflect digesta transit through some component of the colon and rectum. The "fast compartment" or rate constant associated with the ascending phase of the faecal excretion curve can be interpreted as digesta passage through the caecum. However, this interpretation may be very wrong indeed as it is the "fast compartment" which is quantified with time-dependent parameters in the Pond et al. (1988) models. As we have seen above, digesta passage through the caecum seems to be adequately described by time-independent exponential relationships. The "slow compartment" or rate constant associated with the descending phase of the faecal excretion curve can be interpreted as digesta passage through the colon. However, this simplistic interpretation may again prove to be wrong since slower passage rates have sometimes been associated with the time-dependent "ascending phase" or so called "fast compartment" in some equine data sets (Moore-Colyer et al., 2003). Biological interpretation of marker excretion curves and digesta passage in equines seems set to be an interesting area of research over the coming decade.

Figure 12.8.
Time-independent digesta passage models fitted to faecal excretion data for Yb marked fibre feedstuff.

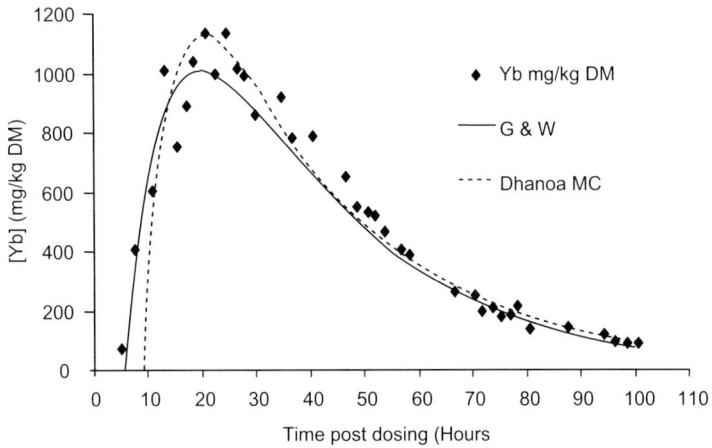

Figure 12.9.
Time-independent digesta passage model (G3G1) fitted to faecal excretion data for Dy marked concentrate feedstuff.

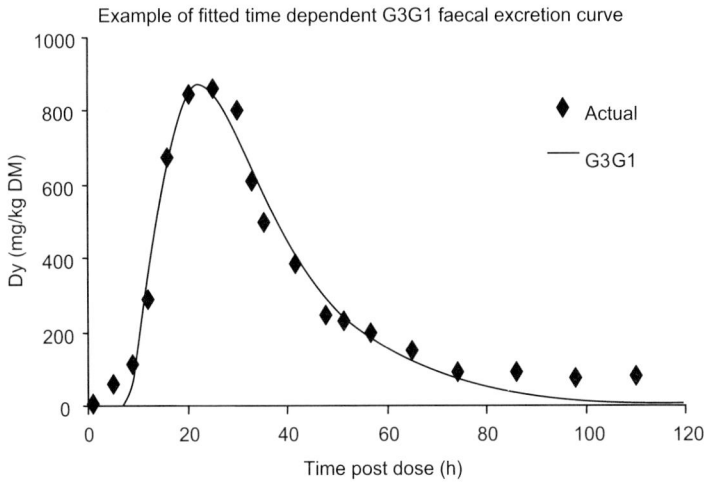

Example of fitted time dependent G3G1 faecal excretion curve

Integrating feed degradation profiles with digesta passage rate estimates

Combining the fitted mathematical parameters from these dynamic feed degradation models with estimates of digesta passage through the pre-caecal, caecal and colonic segments of the equine digestive tract, allows the extent of feed degradation (ED) within each of these major segments to be quantified and compared across a wide range of feedstuffs. Whilst space precludes a detailed explanation of these various models in this paper, ED values for segments where digesta passage is time-independent can be calculated as:

105

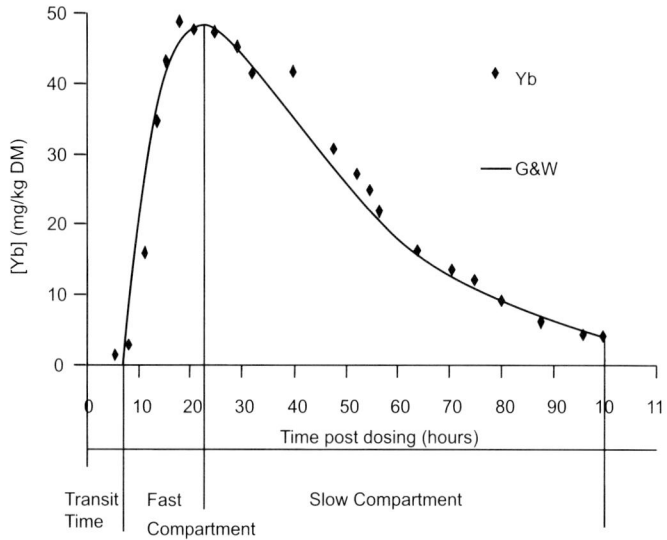

Figure 12.10.
A simplistic initial
description of
possible digesta
passage
compartments in
relation to equine
faecal marker
excretion curves.

$ED = a + (b \times c)/(c + k)$ where a, b and c are as defined by Ørskov and McDonald (1979) as follows:

a = the readily soluble or extractable feed portion and represents the zero time measure of degradability
b = the potentially degradable feed portion
c = the fractional rate of degradation (%/h) of the potentially degradable feed portion (b)

and k is fractional rate of digesta passage (%/h) determined as the exponential rate constant from time-independent declines in marker concentrations following delivery of a pulse dose of marked feedstuff

Similarly, assuming a Gamma 4 time-dependent function (Ellis *et al.,* 1994), ED values for segments where digesta passage is time-dependent can be calculated as:

$ED = a + (b \times c)/(c + ((4/MRT) \times 0.40857))$ where a, b and c are as defined above

and where MRT is mean residence time of feedstuff within the digestive tract segment under study.

To illustrate the principles of how fitted degradation profiles, knowledge and mathematical parameterisation of digesta passage and calculations of effective degradability allow us to partition the digestion of feedstuffs along the equine digestive tract the following working assumptions have been made:

- small intestinal passage is time dependent with a MRT of 3 hours in the pre-caecal segment of the tract
- caecal digesta passage is time independent with a MRT of 4 hours in the caecum
- passage of digesta through the colon is time independent with a MRT of 30 hours.

Data was combined from a series of experiments where unmolassed sugar beet pulp (USBP), ground and pelleted hay cubes (HC), soya hulls (SH) and a 2:1 mixture of oat hulls and naked oats (OHNO) were evaluated using the in vivo experimental and mathematical modelling approaches described above. Dry matter ED values (g/kg), partitioned into the amounts degraded in the small intestine, caecum and colon for each of these 4 feeds are summarised in Table 12.1. The results show that ED of different individual feeds can vary considerable within the same segment of the digestive tract (e.g. ED in the caecum varied from 37 (OHNO) – 411 (USBP) g/kg across feeds). Similarly, the results also demonstrate the wide variation in feed degradability between different segments of the digestive tract for the same feed (e.g. ED for OHNO varied from 11 (colon) – 344 (small intestine) g/kg across segments). A graphical illustration of the dynamic nature of these feed degradation profiles in different segments of the digestive tract along with the calculated ED values for USBP and HC are given in Figure 12.11.

Table 12.1. Partitioned dry matter ED values in the small intestine, caecum and colon of ponies in four equine feeds.

Digestive tract segment	USBP	HC	SH	OHNO	s.e.
Small intestine (i.e. pre-caecal segment)	160	312	246	344	40.7
Caecum	411	124	66	37	85.8
Colon	90	40	177	11	36.4

g/kg

Experimental adaptation of the mobile bag and in situ porous bag techniques for use in nutrition studies with equines described in this paper has allowed the quantitative partition of feed degradation amongst the major segments of the equine digestive tract. However, more research is urgently needed in relation to the dynamics of digesta passage through the various segments of the equine digestive tract. Further studies are now required to capitalise on this development and partition degradation of individual feed constituents in common equine feeds across each of the major segments of the equine digestive tract. If this were done, then diets could be formulated to provide the optimum amounts of degradable feed constituents (sugars, starches, fibre and protein) in each of the major digestive tract segments. This would allow optimum, balanced and healthy diets to be formulated in practice across a wide range of equine husbandry systems.

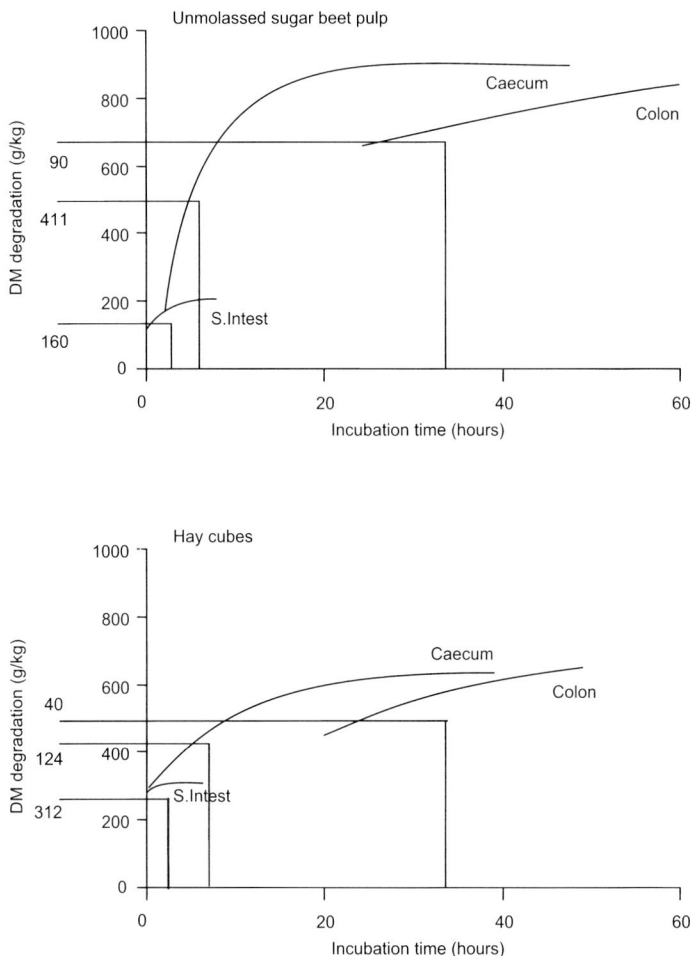

Figure 12.11. Feed degradation profiles in the small intestine, caecum and colon of equines for unmolassed sugar beet pulp and hay cubes along with partitioned effective degradability values calculated for each segment.

These porous bag methods also have a number of experimental uses. Hyslop *et al.* (1998) have used the mobile bag technique to study the dynamics of dry matter and fibre degradation in four forage feedstuffs over the whole digestive tract using intact ponies. This use of mobile bags is more widely applicable than the uses described above since it does not require the use of fistulated animals.

One other key area of research that would benefit from use of this methodology is the partitioning of protein digestion in the horse. It is widely accepted that horses derive their amino acid needs from protein digested in and absorbed from the pre-caecal segment of the digestive tract. However, partitioning protein digestion into that which occurs pre-caecally and that which occurs in the hindgut (caecum and colon) has been difficult in the past. The techniques described here offer one quantitative methodology that allows *in vivo* measurement of pre-caecal protein digestion in equines on a comparable basis across a wide range of both forage and concentrate feed types.

Once quantitative small intestinal protein availability was defined *in vivo*, then *in vitro* methods of prediction could be developed. Using these *in vitro* predictive models, routine assessments of customer specific feed samples could be characterised, small intestinal protein availability defined and the data used in equine rationing solutions tailored to the specific customer/horse requirements. This would allow more accurate construction of diets that supplied the correct type of protein in the correct amounts to help maximise feed and cost efficiency, minimise digestive disorders and alleviate other health problems.

Quantifying the relative degradation of starch and fibre in different segments of the equine tract could also be undertaken to help fine-tune diets for specific equine athletes (e.g. long distance event horses). Finally, further examination of the wide range of feed processing methods employed by the equine feed industry could also be undertaken using the methods described here. Both starch and fibre degradation could be studied in the different segments of the equine tract and the effects of processing methods and treatments (heat, enzymatic, microbial) examined.

Acknowledgements

The UK Horserace Betting Levy Board supported this work.

References

Applegate, C. S. and Herschberger, T. V. (1969). Evaluation of *in vitro* and *in vivo* caecal fermentation techniques for estimating the nutritive value of forages for equine. *Journal of Animal Science.* **28**: 18-22.

Cottrell, D. F., Jones, A. J. and Potter, K. E. (1998). Gas handling in the caecum of the horse. *Experimental Physiology.* **83**: 397-408.

Dhanoa, M.S. (1988). On the analysis of dacron bag data for low degradability feeds. *Grass and Forage Science.* **43**: 441-444.

Dhanoa, M. S., Siddons, R. C., France, J. and Gale, D. L. (1985). A multicompartmental model to describe marker excretion patterns in ruminant faeces. *British Journal of Nutrition.* **53**: 663-671.

Drogoul, C., Faurie, F. and Tisserand, J. L. (1995). Estimation of the contribution of the pony's colon in fibre digestion: a methodological approach. *Annales de Zootechnie.* **44 (Supp)**: 182.

Ellis, W.C., Matis, J.H., Hill, T.M. and Murphy, M. R. (1994). Methodology for estimating digestion and passage kinetics of forages. In: *Forage Quality, Evaluation and Utilisation*. Edited by G.C. Fahey, M. Collins, D.R. Mertens and L.E. Moser. American Society of Agronomy, Madison, WI, USA. pp. 682-756.

Grovum, W. L. and Williams, V. J. (1973). Rate of passage of digesta in sheep. 4. Passage of marker through the alimentary tract and the biological relevance of rate-constants derived from the changes in concentration of marker in faeces. *British Journal of Nutrition*. 30: 313-329.

Hvelplund, T. (1985). Digestibility of rumen microbial protein and undegraded dietary protein estimated in the small intestine of sheep and by *in sacco* procedure. *Acta Agriculturae Scandinavica*. 25 (Supp): 132-144.

Hyslop, J.J., McLean, B.M.L., Moore-Colyer, M.J.S., Longland, A.C., Cuddeford, D. and Hollands, T. (1999a). Measurement of caecal outflow rate in ponies using Chromium mordanted feeds. *Proceedings of the British Society of Animal Science, BSAS, Penicuik, Midlothian, UK*. pp. 141.

Hyslop, J.J., Stefansdottir, G.J., McLean, B.M.L., Longland, A.C. and Cuddeford, D. (1999b). *In situ* incubation sequence and its effect on degradation of feed components when measured in the caecum of ponies. *Animal Science*. 69: 147-156.

Hyslop, J.J., Tomlinson, A. L., Bayley, A. and Cuddeford, D. (1998). Development of the mobile bag technique to study the degradation dynamics of forage feed constituents in the whole digestive tract of equids. *Proceedings of the British Society of Animal Science, BSAS, Penicuik, Midlothian, UK*. pp. 129.

Macheboeuf, D., Poncet, C., Jestin, M. and Martin-Rossett, W. (1996). Use of a mobile nylon bag technique with caecum fistulated horses as an alternative method for estimating precaecal and total tract nitrogen digestibility. *Proceedings of the European Association of Animal Production 47th Annual Meeting - Lillihammer, Norway*. pp. 297.

McLean, B.M.L. (2001). Methodologies to determine digestion of starch in ponies. PhD Thesis. University of Edinburgh, Scotland.

McLean, B.M.L., Hyslop, J.J., Longland, A.C., Cuddeford, D. and Hollands, T. (1999). Development of the mobile bag technique to determine the degradation kinetics of purified starch sources in the pre-caecal segment of the equine digestive tract. *Proceedings of the British Society of Animal Science, BSAS, Penicuik, Midlothian, UK*. pp. 138.

Mehrez, A.Z. and Ørskov, E.R. (1977). A study of the artificial fibre bag technique for determining the digestibility of feeds in the rumen. *Journal of Agricultural Science (Cambridge)*. 88: 645–650.

Miraglia, N., Martin-Rosset, W. and Tisserand, J. L. (1988). Mesure de la digestibilite des fourrages destines aux chevaux par la technique des sacs de nylon. *Annales de Zootechnie*. **37**: 13-20.

Moore-Colyer, M.J.S. (2000). Studies on the degradation kinetics of botanically diverse fibrous feeds and their apparent digestibility and rate of passage when fed to ponies. PhD Thesis, University of Edinburgh, Scotland.

Moore-Colyer, M.J.S., Hyslop, J.J., Longland, A.C. and Cuddeford, D. (2002). The mobile bag technique as a method for determining the degradation of four botanically diverse fibrous feedstuffs in the small intestine and total digestive tract of ponies. *British Journal of Nutrition*. **88**: 729-740.

Moore-Colyer, M.J.S., Morrow, H. J. and Longland, A.C. (2003). Mathematical modelling of digesta passage rate, mean retention time and *in vivo* apparent digestibility of two different lengths of hay and big-bale grass silage in ponies. *British Journal of Nutrition*. **90**: 109-118.

Ørskov, E.R. and McDonald, I. (1979). The estimation of protein degradability in the rumen from incubation measurements weighted according to rate of passage. *Journal of Agricultural Science (Cambridge)*. **92**: 499–503.

Pond, K. R., Ellis, W. C., Matis, J. H., Ferreiro, H. M. and Sutton, J. D. (1988). Compartment models for estimating attributes of digesta flow in cattle. *British Journal of Nutrition*. **60**: 571-595.

Sauer, W.C., Jorgenson, H. and Berzins, R. (1983). A modified nylon bag technique for determining apparent digestibilities of protein in feedstuffs for pigs. *Canadian Journal of Animal Science*. **63**: 233-237.

13

Exploiting dietary fibre in equid diets

Meriel J. S. Moore-Colyer [1] and Annette Longland [2]
[1]*Institute of Rural Studies, University of Wales Aberystwyth, Llanbadarn Fawr, Aberystwyth, SY 23 3 AL Wales UK*
[2] *Institute of Grassland and Environmental Research, Plas Gogerddan, Aberystwyth, SY 23 3EB, Wales, UK*

Abstract

Four *in vivo* apparent digestibility trials were carried out over a period of two years using 4 to 6 Welsh cross pony geldings of approximately 300kg LW. In all experiments ponies were individually housed in 10x12 ft loose-boxes with rubber mat flooring. In Experiment 1 three diets were offered *ad libitum* to 6 ponies in a replicate 3 x 3 Latin squares design. In Experiments 2, 3 & 4, four ponies were offered 1.5 – 1.75 % of LW per day in dry matter (DM) in 4 x 4 Latin square changeover designs. In all experiments periods lasted for 21 days, a 16-day adaptation period followed by a 5-day data collection phase. Diets were fed in two equal meals per day and water was available *ad libitum*. Dried feed and faecal samples were analysed for DM, ADF, NDF and CP using the methods of the AOAC (1990), GE was determined via bomb calorimetry and non-starch polysaccharides (NSP) were quantified by the method of Englyst and Cummings (1984). Differences between diets were determined by analysis of variance (Genstat 5, 2000). NSP analysis gave detailed information on the individual fibre constituents and allowed the subtle differences between feeds to be identified. The DMI of grass and red clover silages were 0.6 and 1.7 kg/d greater than hay indicating that the low DM content of these two forages does not negatively affect feed intake. The apparent digestibilities of these two feeds were also significantly greater than hay for all parameters measured. In Experiment 2 the combination diets of alfa-beet and hay (AB:H) and sugar beet and hay (SB:H) fed in 60:40 and 70:30 ratios were all significantly more digested at > 50% than the hay only diet at <39% for DM, CP, NDF, ADF, NSP, Ca, P and Mg. The addition of AB to H in the 70:30 diet produced a DE value of 2.89 MJ/kg DM higher than the hay alone only diet and also caused an associative affect which improved the digestibility of CP, arabinose, galactose and uronic acids from the hay portion of the diet. No significant differences were noted between the digestibilities of the proximate constituents for the chopped hay replacer (CHR), CHR and highly digestible fibre (HDF) cubes and CHR and course mix (CM) fed in Experiment 3. When looking at the digestibilities of the individual NSP constituents the CHR : CM diet was generally less digestible than the CHR: HDF diets indicating that HDF cubes are equally well or more digestible than a diet containing conventional course mix. With DE values similar to the CM the HDF feed can be offered as a valuable alternative to concentrates in equid diets. The high fibre forage (HiFi) fed in Experiment 4 was also highly digestible for all constituents measured and showed that a combination of chopped straw and alfalfa makes an excellent fibre feed for horses. The results from the four experiments reported here show that detailed chemical analysis is extremely valuable when choosing complimentary fibre feeds for horses. Moreover all the fibre feeds examined here were more digestible than hay. Feeds like the HDF cubes can be used to replace some or all of the cereal portion of the diet. The associative effects between the alfa-beet and hay and the mineral digestibility from these alternative feeds also suggests that not only is gut health enhanced, but that the supply of nutrients from these feeds makes them ideal choices for stabled horses.

Introduction

In order to maximise digestive efficiency and avoid diet related metabolic disorders, it is important to keep sight of the fact that the normal rate of passage of digesta through a horses gut is approximately 28 – 38 hours (Moore-Colyer, 2000). This means that under normal grazing conditions, and for stabled horses fed *ad libitum* forage, the average consumption of feed in dry matter (DM) per day ranges from 2 – 2.5% of LW (Lawrence, 1994). Therefore for a horse to be occupied, feel satiated and consume sufficient nutrients to meet it's daily requirements, whether they are for maintenance or hard work, the diet needs to contain enough forage that will allow the animal to spend a high proportion of its time eating, while contributing significantly to the nutrient density of the diet. All too often in an attempt to increase nutrient intake, horses are fed high concentrate low fibre diets, which not only pre-dispose these animals to diet related metabolic disorders but can also lead to crib-biting, weaving, box walking and wind sucking, which are behaviours that are exacerbated by an inability to perform natural foraging behaviour (Mc Call, 1993; Gillham, 1994).

Hay is the most common forage offered to stabled horses in the UK, but unfortunately most UK hays have low energy and protein contents (Frape, 1986) so they are seen as indigestible ballast whose sole purpose is to maintain digesta rate of passage (Wolter, 1993) and to occupy time for the stabled horse, while the cereal component is regarded as the primary nutrient source. These conventional diets need to be re-evaluated if the health and welfare of the horse is to be improved, thus it is worth exploring alternative fibre feeds such as silage, chopped dried grass, alfalfa and high-fibre pellets as partial or complete replacements for hay. Goodson *et al.* (1988), and Moore-Colyer *et al.* (2000) have demonstrated the beneficial effects of a range of high-energy fibrous diets on volatile fatty acid profiles within the equid hindgut. The net yield of adenosine tri-phosphate (ATP) produced from the aerobic catabolism of acetate, propionate and butyrate are 10, 17 and 25 moles respectively (Mc Donald *et al.,* 1996), and although these do not produce as much ATP as the metabolism of glucose (38 moles of ATP per mole of glucose), Glinisky *et al.* (1976) reported that 30% of the energy available to horses from a 2:1 forage:conc diet came from VFA production in the caecum alone. Thus high fibre diets not only maintain healthy VFA profiles but they also increase the total production of VFAs, which is of major importance to the energy balance of the animal.

In order to maximise VFA production, a factor of critical importance to a performance horse, it is essential to choose readily degradable fibre feeds so that all parts of the ration contribute to the energy density of the diet. Selecting an ideal balance of fast and slow release energy, while taking care of the protein, mineral and vitamin requirements can be difficult to do. However, the process can be made easier if it is known what type of fibre each feedstuff contains.

The widely used term 'fibre' does not refer to a single chemical entity, but rather a fraction of the plant, largely consisting of cell wall material. Dietary fibre therefore consists of lignin and polysaccharides (Aman and Graham, 1990). The potentially degradable portion of the fibre is the polysaccharide fraction (which contains only 10 monosaccharides) and is accurately quantified using non-starch polysaccharide (NSP) analysis (Englyst and Cummings, 1984). NSP is composed of the individual neutral monosaccharides, rhamnose, arabinose, xylose, mannose, galactose, glucose and the acidic sugars glucuronic and galacturonic acids collectively known as uronic acids. Detailed information on the content of these sugars and their individual digestibilities affords a greater understanding of the nutritional value of different fibrous feedstuffs.

All plant cell walls consist of three distinct regions, a middle lamella region, which acts as inter-cellular cement, a primary cell wall and a secondary cell wall. The middle lamella region, which is readily degradable, is the first to develop and consists largely of pectins, which are composed of uronic acids and some neutral sugars in branched chain formation. These chains can be linked via Ca^{2+} bridges. The primary cell wall is also readily degradable and consists of cellulose (ß-linked glucose molecules), which form the structural framework of the plant cell wall (Butler and Bailey, 1973) and hemicellulose, which is composed of various complexes of xylose, arabinose and galactose (Åman and Graham, 1990). Secondary cell walls are formed on the inside of primary cell walls and the formation of these is accompanied by the laying down of lignin (an indigestible complex of phenolic compounds). Depending on the plant species and the stage of growth, secondary cell walls may account for up to 90% of cell wall dry matter (Aman and Graham, 1990). Cellulose, the most important component of secondary cell walls, is formed into microfibrils that are linked by hydrogen bonds and laid down in a highly ordered pattern of three distinct layers (Reid, 1997). The association with lignin makes secondary cell walls hydrophobic (Birch and Parker, 1983) and can render much of the cellulose unavailable to hindgut micro-organisms.

As the formation of cell walls follows the above pattern of middle lamella, primary cell wall and then secondary cell wall, the degradability of fibre decreases with increasing maturity, thus silage, which is cut at an earlier stage of growth, is more digestible than more mature hay (Moore-Colyer and Longland, 2000). However, the above is a generic description of cell wall formation and different plant species have different proportions, complexes and linkages between the different monosaccharides and this can significantly affect fibre digestibility. Thus the only way to determine the extent of digestibility of different feeds in horses is to perform detailed chemical analysis on the feeds and measure the digestibility of the individual components in *in vivo* apparent digestibility trials.

In vivo trials on grass and red clover silages in big bale form, have shown that these feeds are palatable and highly digestible by horses with energy values of 1.5 to 2 times that of hay (Moore-Colyer and Longland, 2000; Hale and Moore-Colyer, 2001), however, despite these findings these feeds are still under utilised as many owners are worried about the safety aspects of feeding silage. Red clover being a legume, has a higher proportion of pectins and minerals than grass and is aerobically stable, with open bales remaining fresh for up to 10 days (Dewhurst, pers com), thus red clover silage can be a valuable fibre to feed to performance horses. Goodwin *et al.* (2002) have looked at the effect of offering multiple forages to stabled horses and reported that most animals preferred forages other than hay. Moreover, when a variety of forages were offered to stabled horses they spent more time eating and less time performing aberrant behaviours.

The work to-date would therefore seem to suggest that the digestive health, energy intake and natural foraging behaviour of the horse are all enhanced when stabled horses are offered a variety of forages. Thus, the objectives of the studies described below were to measure the dry matter intake and digestibility of a wide range of fibrous feedstuffs in order to determine their suitability as partial or complete replacements for hay in equid diets.

Materials and methods

A series of 4 *in vivo* apparent digestibility trials were conducted using Welsh cross pony geldings (*ca.* 300kg LW) fed a variety of fibre-based diets. At the start of each experiment all the ponies were wormed using ivermectin (Furexel, Janssen Animal Health) according to the manufacturer's recommendations and had their feet trimmed by the farrier. In all experiments the ponies were individually housed in 10 x 12 ft loose-boxes with rubber mat flooring. During the 16-day adaptation phases the ponies were allowed approx 1 hour of free exercise per day in a sand arena, while during the 5-day data collection period the ponies were hand walked for 0.5 hours/day. In Experiment 1 three diets were offered *ad libitum* to 6 ponies in replicate 3 x 3 Latin squares designs, ensuring a minimum of 5% of the daily intake was collected as feed refusals. In Experiments 2, 3 & 4, ponies were offered 1.5 – 1.75 % of LW per day in dry matter (DM) in 4 x 4 Latin square changeover designs. In all experiments periods lasted for 21 days, a 16-day adaptation period followed by a 5-day data collection phase. Diets were fed in two equal meals per day and water was available *ad libitum*. Dried feed and faecal samples were analysed for DM, ADF, NDF and CP using the methods of the AOAC (1990), GE was determined via bomb calorimetry and non-starch polysaccharides (NSP) were quantified by the method of Englyst and Cummings (1984). Differences between diets were determined by analysis of variance (Genstat 5, 2000).

Results

Feed composition

The nutrient contents of the feeds offered in Experiment 1 are detailed in Table 13.1. The nutrient compositions of the three forages reflect the botanical origin and stage of maturity at conservation, with the fermented forages having lower DM, ADF and NDF and higher CP contents than the H. The low water-soluble carbohydrate (WSC) content of the RCS is due to the fact that the principal storage carbohydrate in legumes is starch, whereas that in grasses (*graminaceous spp*) is WSC.

In contrast to the silages offered to the ponies in Experiment 1, the feeds fed in Experiments 2, 3 and 4, (Table 13.2) with the exception of the hay were all chopped or ground fibrous feeds that are sold as a 'value added products' and are guaranteed to be of uniform quality and to be free of soil and moulds. The CHR and HiFi are generally marketed as partial hay replacers, whereas the HDF cubes are designed to replace the cereal portion of the diet. The AB and SB are generally fed with the concentrate portion of the ration and are highly palatable and nutritious feeds and contribute to the fibre intake of the diet.

Table 13.1.
Nutrient Content (g/kg DM) of hay (H) grass silage (GS) and Red Clover silage (RCS) offered *ad libitum* to ponies in two 3 x 3 Latin Square designed experiments.

Nutrient	H	GS	RCS
DM	852	371	268
OM	919	916	866
CP	74	104	194
ADF	417	349	331
NDF	723	523	415
GE (MJ/kg DM)	18.9	21.0	19.6
WSC	103	110	8.9

The ADF, NDF CP and TNSP contents detailed in Table 13.2 reflect the stage of maturity of the grasses and legumes at harvest and the species they contain. The alfa beet (AB), for example has very high CP, low ADF and NDF contents compared with the other fibre feeds, reflecting the alfalfa content in this feed. All the processed feeds also have notably higher CP contents than the H, suggesting that these feeds alone would be adequate to meet the maintenance needs of non-working mature equids, whereas the H at only 7% may be deficient in this nutrient.

The individual NSP profiles vary between grass and legume / dicotyledonous based feeds with the AB, SB and HDF containing significant amounts of arabinose, glucose and uronic acids, whereas the H, CHR and HiFi contain notably more xylose and glucose with little arabinose or uronic acid. Mineral contents of the AB, SB and HiFi are similar with between 10 and 19 g/kg DM of Ca and approximately 2 g/kg DM of both Mg and P. The Mg and P contents of the H are similar to the other feeds although the level of Ca is lower at only 6.2 g/kg DM.

Dry matter intake and apparent digestibility

The DMI of RCS (Table 13.3) was 1.1 kg per day higher than for GS and significantly ($P<0.05$) greater than for hay indicating that the RCS fed in this experiment was highly palatable. The DM content of the H, GS and RCS was 840, 370 and 270 g/kg respectively, which meant that the ponies were consuming 6.4 kg of H, 16.4 kg of GS and 26.6 kg of RCS per day on an as fed basis. This demonstrates that gut capacity is not a constraining factor on voluntary food intake when ponies are fed low DM forages.

Table 13.2. Chemical composition of hay (H), alfa beet (AB) and sugar beet pulp (SBP) offered to 4 ponies in Experiment 2; chopped hay replacer (CHR), highly digestible fibre pellets (HDF) and course mix (CM) offered to 4 ponies in Experiment 3 and HiFi, main ring (MR) and XP Yeast, offered to 4 ponies in Experiment 4.

Nutrient	Experiment 2			Experiment 3			Experiment 4		
	H	AB	SB	CHR	HDF	CM	HiFi	MR	XP yeast
DM	810	950	910	870	857	837	800	880	920
OM	947	910	945	896	912	908	905	932	954
ADF	412	249	256	364	242	135	405	234	64
NDF	706	474	585	636	450	328	629	449	260
CP	70	193	108	160	176	154	107	111	152
GE (MJ/ Kg DM)	17	18	^	17.8	17.1	16.6	^	^	^
Starch	^	^	^	^	^	^	38	258	383
TNSP	463	406	514	302	306	248	437	323	198
Rhamnose	0	7	12	1	2	1	2	1	6
Arabinose	28	88	168	20	40	28	25	31	40
Xylose	138	22	11	77	80	76	100	104	59
Mannose	2	9	7	2	12	2	9	3	6
Galactose	10	26	41	8	10	5	11	8	9
Glucose	270	146	131	185	145	130	254	153	70
Uronic acids	16	108	144	9	17	6	36	18	9
Ca	6.2	18.7	12.2	^	^	^	10	10	2
Mg	1.3	2.7	1.8	^	^	^	2	2	4
P	1.7	2.3	0.1	^	^	^	2	6	8

^ Values unavailable

Table 13.3. Dry matter intakes (kg/d), in vivo apparent digestibility coefficients and digestible energy (DE) contents (MJ/kg DM) of hay (H) big bale grass silage (GS) and red clover silage (RCS) by ponies.

	H	GS	RCS	s.e.d
DMI	5.5[a]	6.1[ab]	7.2[b]	0.44
DMD	0.36[a]	0.69[b]	0.74[c]	0.012
CPD	0.29[a]	0.69[b]	0.81[b]	0.051
NDFD	0.32[a]	0.64[b]	0.66[b]	0.014
ADFD	0.28[a]	0.63[b]	0.68[b]	0.010
DE	6.0[a]	14.5[b]	14.4[b]	0.16

[abc] Values in the same row with different superscripts differ significantly (P<0.05)

The apparent digestibilities of GS and RCS were significantly greater than H for all the parameters measured. The higher DMD and ADFD noted for RCS compared with GS, reflects the higher content of readily degradable material (CP being 193 g/kg DM) and the lower cell wall content of 331 and 415 g/kg DM for ADF and NDF respectively for the RCS compared with the 104 CP 349 ADF and 523 NDF g/kg DM noted in the GS. Digestible energy (DE) contents of GS and RCS were both greater than 14 MJ/kg DM and significantly higher than that noted for H at only 6 MJ/kg DM.

The DMIs detailed in Table 13.4 from Experiment 2 show that a 60:40 ratio of AB:H resulted in a significantly lower feed intake than when H was fed with 30% AB or SB. Digestibility of all three-combination diets at >50%, for DMD, OMD, CPD, NDFD and NSPD, were significantly greater (P<0.05) than for H at <39%. The digestibility coefficients of AB and SB calculated by difference show all values to be between 0.57 and 1. The 1.02 of CP digestibility for AB (30%) suggests an associative effect between the AB and the H with an enhanced digestibility of CP from the hay portion of the diet. This associative effect is also noted for the individual NSP constituents arabinose, galactose and uronic acids. Thus it appears that both AB and SB enhance the digestibility of both the CP and the fibre fraction of H.

	H	AB:H	H:AB	H:SB	s.e.d
DMI	4.4[b]	3.9[a] (2.3)	4.3[b] (1.3)	4.5[b] (1.3)	0.16
DMD	0.38[a]	0.59[c] (0.74)	0.50[b] (0.81)	0.54[bc] (0.89)	0.034 (0.090)
CPD	0.37[a]	0.66[c] (0.85)	0.56[bc] (1.02)	0.52[b] (0.81)	0.054 (0.136)
NDFD	0.32[a]	0.52[b] (0.65)	0.51[b] (0.96)	0.49[b] (0.84)	0.049 (0.201)
ADFD	0.25[a]	0.43[b] (0.56)	0.35[ab] (0.60)	0.40[b] (0.71)	0.048 (0.127)
NSPD	0.39[a]	0.63[b] (0.79)	0.50[ab] (0.75)	0.58[b] (0.98)	0.064 (0.097)
Rhamnose D	1	1 (1)	1 (1)	0.98 (1)	0.010 (0.041)
Arabinose D	0.53[a]	0.89[c] (1.14[a])	0.78[b] (1.38[b])	0.86[c] (1.55[c])	0.034 (0.053)
Xylose D	0.45	0.32 (0.23)	0.32 (-0.03)	0.32 (0.07)	0.156 (0.24)
Mannose D	0.89	0.83 (0.77)	0.82 (0.66)	0.86 (0.83)	0.099 (0.149)
Galactose D	0.58[a]	0.85[c] (1.03[a])	0.76[b] (1.19[b])	0.82[bc] (1.31[c])	0.028 (0.041)
Glucose D	0.33[a]	0.52[b] (0.66)	0.44[ab] (0.71)	0.49[b] (0.83)	0.051 (0.090)
Uronic acid D	0.49[a]	0.90[c] (1.19[a])	0.82[b] (1.61[b])	0.90[c] (1.74[b])	0.023 (0.079)
CaD	0.56[a]	0.69[c] (0.77[ab])	0.65[bc] (0.81[a])	0.58[ab] (0.61[b])	0.036 (0.072)
MgD	0.43[a]	0.60[b] (0.72)	0.56[b] (0.87)	0.52[ab] (0.68)	0.041 (0.127)
PD	0.27	0.33 (0.36)	0.31 (0.39)	0.28 (0.30)	0.118 (0.367)
Nutritive value					
DCP content	26.4[a]	84.2[d]	60.0[c]	42.2[b]	6.58
GED	317	^	458	^	^
DE (MJ/ kg DM)	4.96	^	7.85	^	

Table 13.4. Dry matter intakes (kg/d), in vivo apparent digestibility coefficients and digestible energy (DE) contents (MJ/ kg DM) of three forage based diets, Hay (H), Alfa-beet : H (AB:H) 60:40 ratio, H:AB 70:30 ratio and hay: sugar beet pulp (H:SB) by ponies.

Values in italics and () are intake and apparent digestibility values for AB and SB alone, calculated by difference.
[abc] values within the same row with different superscripts differ significantly (P<0.05).

In Experiment 3 (Table 13.5), there were no significant differences in digestibility between the four different combination diets for DM, CP, NDF and ADF suggesting that HDF pellets are equally well digested as conventional course mix when fed in a 50:50 ratio with CHR. However, the addition of the FOS to the CHR:HDF diet resulted in slightly higher digestibilities of all the individual as well as the total NSPs than when the FOS was omitted as in diet 2. This made diet 3 significantly more digestible than CHR:CM and CHR alone. The addition of FOS also significantly increased the DE content of diet 3 making the DE content of 9.9 MJ/kg equal to that of the CHR:CM diet and significantly greater than the CHR:HDF diet. No significant differences between diets for any of the parameters measured were noted in Experiment 4, indicating that HiFi is equally well digested as the combination diets of HiFi and course mix. The addition of VXP yeast did marginally increase the digestibility of most of the parameters measured, although none of these differences were significant.

Discussion

Feed composition

The botanically diverse feeds tested in the four experiments reported here indicate that a wide range of plant material can be successfully fed to horses. However, the extent of their individual digestibility does reflect the chemical composition of each plant within each feedstuff, thus to better understand why one feed is more digestible than another, the profile of the individual sugar monomers present were studied. The cell wall composition of the members of the *graminea* family (grasses) can be reasonably accurately determined using NDF analysis as these plants have little or no uronic acids, however legumes and dicotyledons can contain significant amounts of these highly digestible constituents (Longland *et al.*, 1994) and so they need to be accurately quantified using NSP analysis. The fact that the SBP was more digestible than H in Experiment 2 (Table 4), is due to its high content of primary cell wall material, denoted by high glucose, uronic acid, arabinose and low xylose. Likewise, the higher digestibility of the legume, RCS, in Experiment 1 and the HiFi in Experiment 4 compared with H, can be explained by a more detailed examination of the individual NSP contents and their subsequent *in vivo* digestibilities. Thus the results from all the experiments reported here can be accurately explained using the information obtained from the NSP analysis, which shows that the uronic acid fraction in all the feeds is highly digestible (Tables 13.4 and 13.5).

Feed intake

Intakes of the ensiled forages, GS and RCS in Experiment 1

Table 13.5. Dry matter intakes (kg/g), *in vivo* apparent digestibility values and digestible energy (DE) contents (MJ/kg DM) in ponies offered chopped hay replacer (CHR), highly digestible fibre pellets (HDF) and course-mix (CM) in Experiment 3 and HiFi and course-mix (CM2) plus V XP Yeast in Experiment 4.

	Experiment 3					Experiment 4				
	D1	D2	D3	D4	s.e.d	HF	HF+	HF:CM2	HR:CM2+	s.e.d
DMI	3.4	3.5	3.5	3.5	0.08	5.3	5.4	5.2	5.3	^
DMD	0.60	0.56	0.57	0.51	0.043	0.54	0.54	0.56	0.58	0.039
CPD	0.71	0.70	0.71	0.67	0.037	0.66	0.66	0.69	0.71	0.037
NDFD	0.48	0.47	0.48	0.52	0.054	0.42	0.46	0.37	0.38	0.051
ADFD	0.35	0.41	0.42	0.37	0.057	0.40	0.43	0.34	0.35	0.058
NSP	0.54[a]	0.63[ab]	0.73[b]	0.51[a]	0.079	0.56	0.58	0.51	0.55	0.037
Rhamnose D	0.37[a]	0.69[b]	0.77[b]	0.50[ab]	0.111	-	-	-	-	0
Arabinose D	0.57	0.58	0.70	0.56	0.088	0.57	0.72	0.68	0.70	0.078
Xylose D	0.37[ab]	0.38[ab]	0.54[b]	0.22[a]	0.126	0.42	0.46	0.43	0.45	0.049
Mannose D	0.78[b]	0.91[b]	0.93[b]	0.58[a]	0.074	0.75	0.74	0.57	0.56	0.123
Galactose D	0.69	0.72	0.79	0.68	0.087	0.74	0.78	0.74	0.74	0.027
Glucose D	0.48	0.46	0.63	0.39	0.109	0.56	0.52	0.48	0.48	0.048
Uronic Acid D	0.58[a]	0.67[ab]	0.77[b]	0.62[a]	0.058	0.80	0.84	0.79	0.84	0.060
CaD	^	^	^	^	^	0.66	0.56	0.66	0.52	0.080
MgD	^	^	^	^	^	0.55	0.56	0.51	0.53	0.047
PD	^	^	^	^	^	0.09	0.15	0.30	0.28	0.117
Nutritive value										
DCP content (g/kg DM)	111[b]	118[c]	119[c]	105[a]	2.19	71.2	71.2	75.7	79	4.51
GED	575	548	562	482	47.8	^	^	^	^	^
DE (MJ/kg DM)	9.88[c]	9.56[b]	9.94[c]	8.56[b]	0.092	^	^	^	^	^

D1 = 50:50 CHR:CM; D2 = 50:50 CHR:HDF D3 = 50:50 CHR:HDF + 15g fructo-oligosaccharide (FOS) D4 = CHR.
[abc] values within the same row and experiment with different superscripts differ significantly (P<0.05); ^ Values not available.

demonstrate that these feeds are highly palatable and readily eaten by ponies. This agrees with the previous findings of Moore-Colyer and Longland (2000) and Moore-Colyer et al. (2003) who reported fresh weight intakes in excess of 11 kg /d for big bale grass silage. These values clearly show that relatively 'wet' forages do not affect voluntary food intake in ponies. In fact the ponies in Experiment 1 consumed approximately 8% of their live weight per day of RCS and 5.5% of GS indicating that gut capacity is not a limiting factor to the intake of ensiled forages in ponies.

Apparent digestibility

The apparent digestibility values obtained for the range of fibrous feeds tested in these experiments show them to be equal to or greater than a range of different hay fodders tested by several different researchers (Vander Noot and Gilbreath, 1970; Pearson et al., 1991; Cymbaluk 1990; Aiken et al., 1989; Croizer et al., 1997). Additionally the high DE content of some of the feeds tested here were equal to the DE contents of many different course-mix concentrate feeds presently on the market indicating that these fibrous feeds made a valuable contribution to the energy content of the diet and could in future be used to replace hay in performance horse diets. A more energy dense forage should also allow the concentrate portion of the diet to be reduced as the forage would be making a valuable contribution to the nutrient content of the diet thereby allowing less reliance on concentrates. The fibre feeds tested here were a variety of long and chopped grass-based or legume-based forages. These feeds need not be fed in isolation and in fact if several of them were offered together, not only would the nutrient content of the diet be improved but as Goodwin et al. (2002) have reported, the natural foraging behaviour of stabled horses would be enhanced and so aberrant behaviours should be minimized. Moreover, the energy that is derived from fibre feeds is slow-release energy, which in most equestrian sports is infinitely more preferable to the glucose-induced excitability (the glycaemic response) that starch-base cereals produce.

Microbial activity and degradation capacity

Readily degradable fibrous feedstuffs also help to maintain a healthy microbial population as the bacteria, protozoa, fungi and yeasts all have a food source on which to grow and multiply. This improves their degradation capacity and results in a greater proportion of the less degradable fraction of the feed being fermented to VFA, and thus contributing to the DE content of the diet. This associative effect between feeds (concentrates and roughages) has been well documented in ruminants (McDonald et al., 1995), however, Martin-Rossett and Dulphy (1987) found no such affect in equids. It is worth noting however, that these workers were examining the association between concentrates and roughages, which in the horse are digested within different regions of the gut, and thus it is unlikely that any association takes place. But, when looking at the effect one fibrous feed may have on another, the results from the present study and earlier data by Moore-Colyer (2000) show a clear associative effect to be present. This effect is due to the fact that all fibrous feedstuffs are fermented in the hindgut, thus the two feeds examined are within the same chamber so any factor that improves microbial degradation will improve the degradation of all the feeds within that chamber. Fibre feeds such as sugar beet pulp have been shown to improve the degradation of hay *in vitro* (Moore-Colyer, 2001) and the same effect was noted in the first experiment reported here. This may be due to the fact that the CP in sugar beet pulp is held within a hydroxyproline rich fraction within the cell wall and is therefore not released until the feed reaches the hindgut (Moore-Colyer, *et al.*, 2002) thus supplying the microbes with a readily accessible CP source. Moreover, the fibre in sugar beet pulp is composed primarily of pectins (intercellular cement) and primary cell walls, which are readily degraded by the hindgut microbes. Thus the associative effect between sugar beet pulp and hay may be due to a combination of readily degradable CP and an ideal ratio of 2:1 of primary to secondary cell walls, all of which improve microbial growth and synthesis and thus degradation capacity.

Mineral digestibility

The digestibility of Ca, which is held in the pectic fraction of the cell wall, from the sugar beet pulp and alfa-beet ranges from 68 to 87% and supports the generally held view that the availability of Ca is high in sugar beet pulp and alfalfa. In all feeds other than the hay tested in Experiments 2, 3 and 4 the apparent digestibility of Mg was greater than 50% again indicating a high level of availability of this important mineral. P availability from the complete fibre diets fed in Experiment 2 was approximately 30%. In contrast the digestibility of this mineral was notably lower from the HF and HF+ diets in Experiment 4 compared with the combination diets containing cereal-based course mix. When fed in combination with cereals the HiFi balanced the high P content of the course-mix and altered the P digestibility of the total diet to 30%, which was equivalent to that of an all fibre diet. This again demonstrates that the inclusion of significant amounts of fibre in horse diets can re-dress imbalances that are caused by feeding cereal-based concentrates. Moreover, this work highlights the fact that not all fibrous feeds are the same and that detailed chemical analysis is essential if owners want to make informed choices about complimentary feedstuffs in order to ensure that the animal receives a balanced diet.

Feeding a fibre-based diet ensures that the dietary cation-anion balance (DCAB) is good i.e. alkaline in nature. Horses and other grazing animals have evolved on alkaline diets and when subjected to acidic diets (cereals) can suffer from nutritionally induced acidosis. Baker et al. (1993; 1997) have indicated that young horses consuming diets with a low DCAB may be in negative Ca balance due to an increased urinary excretion of calcium. The decrease in blood pCO_2 HCO_3^- and higher concentration of H^+ anions (from the sulphur and chloride in the cereals) drops the pH of the blood (Baker et al., 1992). Low blood pH causes an increase in the levels of ionized Ca in the serum, which is excreted by the kidneys, hence the increased urinary Ca excretion. Increased Ca excretion raises parathyroid hormone (PTH) activity, resulting in increased Ca absorption from the intestines and the bone (Cooper et al., 1998). Although Wall et al. (1997) have demonstrated that weanling horses fed low DCAB were able to compensate for the higher urinary loss by increasing absorption and decreasing faecal output and so maintain Ca balance, this may not be the case for older horses whose absorption of Ca is 20% less that younger animals. Moreover, Wall et al. (1992) demonstrated that anaerobically exercised horses fed low DCAB diets had lower Ca, P and Mg balances, thus older exercising horses may still suffer from negative Ca balances on low DCAB diets. A long-term negative balance of Ca will affect bone development and may be a contributing factor to stress fractures in racehorses. Ensuring good availability of minerals from the diet and reducing conditions that may leach minerals from the body should be a prime consideration when feeding young performance horses, and is another reason to incorporate a higher level of highly digestible fibre into the diet.

Conclusions

The fibre-mix diets offered to ponies in these 4 experiments all proved to be palatable highly digestible feeds, with energy and mineral levels significantly higher than hay. Feeds such as highly digestible fibre pellets and sugar beet pulp can be used to replace some of the concentrate portion of the diet without a loss of energy density while replacing some of the hay with fermented or chopped forages will improve the energy density of the fibre portion of the diet and additionally improve the supply of essential macro minerals. Detailed chemical analysis of feedstuffs improves our understanding of why certain feeds are more degradable than others and should allow owners to choose complementary feedstuffs in order to provide their animals with healthy balanced diets.

Acknowledgements

The authors would like to acknowledge the financial assistance of Dengie Crops Ltd, Spillers Specialty Feeds and Rumenco. We would also like to thank the following MSc students Hermione Ball, Browen Jackson and Catherine Hale without whose help this work would not have been possible.

References

Aiken, G.E., Potter, G.D., Conrad, B.E. and Evans, J.W. (1989). Voluntary intake and digestion of Coastal Bermuda grass hay by yearling and mature horses. *Equine Veterinary Science.* 9: 262-264.

Aman, P. and Graham, H. (1990). Chemical evaluation of polysaccharides in animal foods. In: *Foodstuff Evaluation* Edited by J. Wiseman and D.J. Cole. Butterworths, London, UK. pp. 161-178.

Association of Official Analytical Chemists. (1990). *Official methods of analysis of the Association of Official Analytical Chemists.* 15th edition. Virginia, USA.

Baker, M.S., Topliff, D.R., Freeman, D.W., Teeter, R.G., Breazile, J.E. (1992). Effects of dietary cation-anion balance on acid-base status in horses. *Journal of Equine Veterinary Science.* 12: 160.

Baker, M.S., Wall, D.L., Topliff, D.R., Freeman, D.W., Teeter, R.G., Breazile, J.E. and Wagner, D.G. (1993). Effect of dietary cation-anion balance in anaerobically exercised and sedentary horses. *Journal of Equine Veterinary Science.* 10: 557-561.

Baker, M.S., Topliff, D.R., Freeman, D.W., Teeter, R.G., Stoecker, B. (1997). Comparison of two forms of sodium in the dietary cation-anion difference equation: Effects on mineral balance in sedentary horses. *Proceedings of the 15th Equine Nutrition and Physiology Symposium.* Ft Worth, Texas, USA. ENPS, Savoy, IL, USA.

Birch, G.G. and Parker, K.J. (1983). *Dietary Fibre.* Applied Science Publishers Ltd, London and New York.

Butler, G.W. and Bailey, R.W. (1973). *Chemistry and Biochemistry of Herbage,* Academic Press, London and New York.

Cooper, S.R., Kline, K.H., Foremann, J.H., Brady, H.A. and Frey, L.P. (1998). Effects of dietary cation-anion balance on pH, electrolytes and lactate in standard bred horses. *Journal of Equine Veterinary Science.* 18: 662-666.

Croizer, J.A, Allen, V.G., Jack, N.E., Fontenot, J.P. and Cochrane, M.A. (1997). Digestibility, apparent mineral absorption and voluntary intake by horses fed Alfalfa, Tall Fescue and Caucasian Bluestem. *Journal of Animal Science.* 75: 1651-1658.

Cymbaluk. N.F. (1990). Comparison of forage digestion by cattle and horses. *Canadian Journal of Animal Science.* 70: 601-610.

Englyst, H.N. and Cummings, J.H. (1984). Simplified method for the measurement of total non-starch polysaccharides by gas-liquid chromatography of constituent sugars as alditol acetates, *Analyst.*

109: 937-942.

Frape, D. (1986). *Equine Nutrition and Feeding*, Longman Scientific and Technical, Harlow, Essex, UK.

Gillham, S.B., Dobman, N.D., Shuster, L., Kream, R. and Rand, W. (1994). The effect of diets on cribbing behaviour and plasma ß endorphin in horses. *Applied Animal Behaviour Science*. 41: 147-153.

Glinsky, M.J., Smith, R.M., Spires, H.R. and Davis, C.L. (1976). Measurements of volatile fatty acid production rates in the caecum of the pony. *Journal of Animal Science*. 42: 1465-1470.

Goodwin, D., Davidson, H.P.B. and Harris, P. (2002). Foraging enrichment for stabled horses: effects on behaviour and selection. *Equine Veterinary Journal*. 34: 686-691.

Goodson, J., Tyznik, W.J., Cline, J..H. and Dehority, B.A. (1988). Effects of an abrupt diet change from hay to concentrate on microbial numbers and physical environment in the caecum of the pony. *Applied Environmental Microbiology*. 54: 1946-1950.

Hale, C. and Moore-Colyer, M.J.S. (2001). Voluntary food intakes and apparent digestibilities of hay, big-bale grass silage and red clover silage by ponies. *Proceedings of the 17th Equine Nutrition and Physiology Society*, Lexington, Kentucky, June 2001. ENPS, Savoy, IL, USA.

Laws Agricultural Trust (1993). Genstat 5, Rothamstead Experimental Station, Harpendon, Hertfordshire, UK.

Lawrence, L., (1994). Practical considerations in equine sports medicine. In: *The Athletic Horse* Edited by D.R. Hodgson and R.J. Rose. W.B. Saunders and Co., Philadelphia, USA.

Longland, A.C., Carruthers, J.C. and Low, A.G. (1994). The ability of piglets 4-8 weeks old to digest and perform on diets containing two contrasting sources of non-starch polysaccharide. *Animal Production*. 58: 405-410.

McCall, C.A. (1993). Solving behaviour problems in horses. *Equine Practice*. 15: (8) 30-31.

McDonald,M., Edwards, R.A., Greenhalgh, J.F.D. and Morgan,C. (1995). *Animal Nutrition*. 5th edition, Longman Scientific and Technical, Harlow, UK.

Martin-Rosset, W. and Dulphy, J.P. (1987). Digestibility interactions between forages and concentrates in horses: Influence of feeding level - comparison with sheep. *Livestock Production Science*. 17: 263-276.

Moore-Colyer, M.J.S., Hyslop, J.J., Longland,A.C. and Cuddeford, D. (2000). Intra-caecal fermentation parameters in ponies fed botanically diverse fibre feeds. *Animal Feed Science and Technology*. 84: 183-197.

Moore-Colyer, M.J.S. and Longland, A.C.L. (2000). *In vivo* apparent digestibility of four types of conserved forage by ponies. *Animal Science*. 71: 527-534.

Moore-Colyer, M.J.S. (2000). Studies on the degradation kinetics of botanically diverse fibrous feeds and their apparent digestibility and rate of passage when fed to ponies. *PhD Thesis, University of Edinburgh.*

Moore-Colyer, M.J.S. (2001). The effect of plain sugar beet pulp on the *in vitro* gas production and *in vivo* apparent digestibility of hay when offered to ponies. *Proceedings of the 17th Equine Nutrition and Physiology Symposium* June 2001, Lexington, KY, USA. ENPS, Savoy, IL, USA.

Moore-Colyer, MJ.S., Hyslop, J.J., Longland, A.C. and Cuddeford, D. (2002). The mobile bag technique as a method for determining the degradation of four botanically diverse fibrous feedstuffs in the small intestine and total digestive tract of ponies. *British Journal of Nutrition.* **88**: 729-740.

Moore-Colyer, M.J.S; Morrow, H.J. and Longland, A.C. (2003). Mathematical modelling of digesta passage rate, mean retention time and *in vivo* apparent digestibility of two different lengths of hay and big-bale grass silage in ponies. *British Journal of Nutrition.* **90**: 109-118.

Pearson, R.A. and Merritt, J.B. (1991). Intake, digestion and gastrointestinal transit time in resting donkeys and ponies and exercised donkeys given *ad libitum* hay and straw diets. *Equine Veterinary Journal.* **23**: 339-343.

Reid, G.J.S. (1997). Carbohydrate Metabolism: Structural Carbohydrates. In: *Plant Biochemistry*, Edited by P.M.Dey and J.B. Harborne. Academic Press. London, UK. pp. 205-235.

Vander Noot, G.W. and Gilbreath, E.B. (1970). Comparative digestibility of forages by geldings and steers. *Journal of Animal Science.* **31**: 351-355.

Wall, D.L., Topliff, D.R., Freeman, D.W., Wagner, D.G. and Breazile, J.E. (1992). Effects of dietary cation-anion balance on urinary mineral excretion in exercised horses. *Journal of Equine Veterinary Science* **12**: 168.

Wall, D.L., Topliff, D.R., Freeman, D.W., Breazile, J.E. and Teter, R.G. (1997). The effect of dietary cation-anion balance on mineral balance in growing horses. *Proceedings of the 15th Equine Nutrition and Physiology Symposium.* Ft Worth, Texas, USA. ENPS, Savoy, IL, USA.

Wolter, R. (1993). Fibre in the feeding of horses. *Practique Veterinaire Equine.* **53**: 321-325.

14

Effects of airway inflammation, ozone and exercise on the pulmonary antioxidant capacity of the horse: a war of nutrition

C Deaton
Centre for Equine Studies, Animal Health Trust, Newmarket, Suffolk, UK

Within the body there is continual production of entities known as Reactive Oxygen Species (ROS). These include radical derivatives of oxygen that contain at least one unpaired electron and include species such as the superoxide and hydroxyl radicals. ROS also include non-radical derivatives of oxygen that are capable of oxidising biomolecules such as hydrogen peroxide, ozone and nitrogen dioxide. ROS are formed from processes such as the respiratory burst of phagocytes and from mitochondrial oxidative phosphorylation, so production is often increased by situations that elevate oxygen utilisation such as exercise. ROS may also act as "signalling" species within the body. Controlled production of ROS is therefore essential for normal cellular function and health, especially with respect to the functioning of the immune system. However, uncontrolled production of ROS can result in cell damage and death, the induction and propagation of inflammation and DNA damage. Thus, the body has evolved intricate and elaborate enzymatic and non-enzymatic antioxidant defences to control and buffer excess ROS production. In situations where the antioxidant defences are overwhelmed either due to their depletion, malfunction or simply due to excessive ROS bombardment, oxidative stress and oxidative damage are likely to occur.

ROS have been implicated in many different disease processes, including arthritis, cardiovascular disease and cancer. ROS have also been implicated in the ageing mechanism. However, the lung appears to be particularly susceptible to "oxidative stress" and ROS have been suggested to play an important role in conditions including asthma, ARDS, NRDS, oxygen toxicity, idiopathic pulmonary fibrosis, COPD (human), cystic fibrosis and emphysema. Equine recurrent airway obstruction (RAO, formerly known as equine chronic obstructive pulmonary disease, COPD) is a condition with many similarities to human asthma and has been estimated to be the most common medical condition of horses in Northern Europe. Affected horses develop bronchoconstriction, mucus accumulation and neutrophilic airway inflammation in response to inhaled allergens and ROS and antioxidants have been shown to play an important role in this condition.

We have recently identified ascorbic acid as the major non-enzymatic antioxidant in equine bronchoalveolar lavage fluid (BALF). Ascorbic acid is present in the BALF of the healthy horse at a concentration approximately forty times greater than that of glutathione and twenty times the BALF ascorbic acid concentration of man. The concentrations of ascorbic acid in both plasma and BALF are lower in RAO-affected horses in the absence of inflammation compared to non-RAO-affected horses. In addition, the ratio of oxidised ascorbic acid (dehydroascorbate, DHA) to total ascorbic acid (DHA and reduced ascorbic acid), referred to as the ascorbic acid redox ratio, is greater in the presence of airway inflammation. In control and RAO-affected horses exposed to hay and straw for twenty-four hours, airway inflammation was greater in the RAO-affected and in these horses there was a significant reduction in BALF ascorbic acid concentration.

As RAO-affected horses have lower concentrations of BALF ascorbic acid compared to control horses we hypothesised that RAO-affected horses would be more susceptible to the oxidising effects of ozone due to their lower antioxidant status. Three out of seven non-RAO affected horses and three out of seven RAO-affected horses developed marked increases in oxidised glutathione (GSSG) in BALF in response to ozone, indicative of pulmonary oxidative stress. This suggests that there are responders and non-responders to ozone independent of disease and antioxidant status.

As RAO-affected animals have lower BALF ascorbic acid concentrations, we have investigated the effect of supplementation with two different sources of ascorbic acid. Despite the inherent capacity for ascorbic acid production, the BALF concentration of ascorbic acid increased in five out of six ponies following supplementation.

Exercise has previously been demonstrated to induce oxidative stress therefore the effects of exercise on pulmonary and systemic antioxidant status with and without dietary antioxidant supplementation were investigated in both non-RAO and RAO-affected horses. Interestingly, moderate (up to 90% VO_{2max}) intensity exercise failed to induce oxidative stress in either group. There were no clear beneficial or detrimental effects of supplementation on pulmonary function or indices of oxidative stress or muscle damage during or after exercise and the importance of antioxidant supplementation and increased antioxidant reserves may not become apparent unless the diet is deficient in antioxidants, exercise intensity is higher or more prolonged, unless disease is present or animals are subjected to additional stress such as transport.

This work was funded by The Equine Studies Group of the WALTHAM Centre for Pet Nutrition.

15

Equine forages and grassland management

J Bax[1] and G P F Lane[2]
[1] Biotal Ltd, Collivaud House, Cardiff CF24 5PD, UK
[2] The Royal Agricultural College, Cirencester, Gloucestershire GL7 6JS, UK

Introduction

There are about 900,000 horses kept in the UK for leisure and a further 65,000 kept professionally (British Horse Society, 2003). Altogether about 800,000 ha are utilised for grazing and feeding horses, making the equine industry the second largest in the UK land based sector, after agriculture. The purpose of this paper is to review the present information on the management of grassland for horse grazing and for conserved forage, to identify some current issues and to postulate some areas for future research and development.

Grazing management

Maximizing the contribution made by grazing to the nutrition of the horse is a reasonable goal if economy is the main priority. Even if this is not the case, horses have a need to satisfy their natural instinct to graze. The deterioration of poorly managed grassland associated with equine enterprises is well documented (Odberg and Francis-Smith, 1976, and Judd et al., 2001). Horse grazing differs from grazing by cattle and sheep, in particular in the way that they select species and in the spatial pattern of their grazing. The resultant areas of "roughs" and "lawns" are familiar to all horse keepers and often lead to serious pasture degradation through both over and under-utilisation. Archer (1978) suggested that only 10% of the area of long established horse pastures might be grazed. Gibson (1998) refers to many of the key features of horse grazing. In addition to the above, physical damage resulting from high stocking rates, exercising, and extended periods at grass, as well as the introduction of conserved fodder with its associated weed seeds, may all cause species degradation. Judd et al. (2001) reported on a horse pasture management study in South East England and referred to ways in which horse grazing can lead to detrimental landscape effects through inappropriate management and parallel developments, particularly in areas of countryside close to towns. In many cases horse owners do not have the inherent skills in grassland management which farmers possess and so it is perhaps not surprising that deterioration occurs so frequently.

The main factors which affect the productivity of all grassland are, soil type, nutrient supply and drainage, altitude and climate, and sward species composition. The combination of the effects of soil type, altitude and rainfall are very considerable, to the extent that grassland in high rainfall areas with moisture retentive (silty or clayey) soils can be almost 50% more productive than grassland in dry areas on shallow or sandy soils (Thomas *et al.*, 1991). Horse keepers seldom take note of such important factors when deciding on stocking densities and seem happy to adhere to the "industry standard" of 0.4 ha / horse as the ideal wherever that horse or its grazing is situated. Furthermore, the rate of grass growth changes substantially during the season (see Figure 15.1, ADAS, 1995), necessitating changes in optimum stocking density at different times of year. It is rare to find horse keepers aware of such factors or making management decisions to mitigate them in order to maintain a reasonably constant supply of nutrients to their stock while at grass.

Figure 15.1. Annual curve of the rate of grass growth (ADAS, 1995)

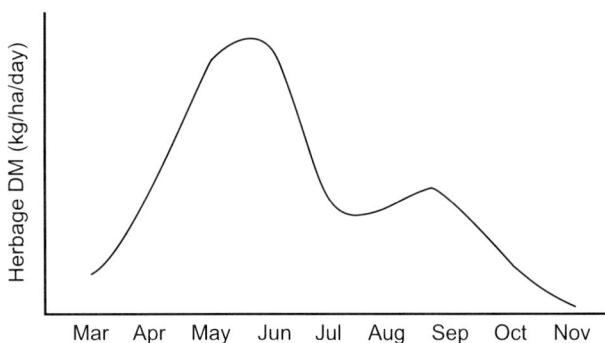

(Source: ADAS)

Soil fertility is obviously an important issue and the requirements for lime, N, P and K on grassland are well understood (MAFF, 2000). However, Judd *et al.* (2001) reported that only 29% of horse keepers in their survey were (or had in the last 20 years) undertaking pH or any other nutrient testing although 71% were using fertilisers to some extent. It is important to maintain soil pH at about 6.5 since this factor alone can substantially enhance sward productivity, especially where levels have been allowed to decline. In the Guide to Classical Experiments carried out at Rothamsted Experimental Station (AFRC, 1991) a response of about 29% to liming, in terms of annual sward dry matter yield, was reported from the "Park Grass" trial, where soil pH was variously maintained at between 5 and 7 over an extended period. Liming with ground chalk or ground limestone is fairly cheap but usually involves the use of contractors equipment so should normally be carried out in late summer or autumn when there is least likelihood of soil or sward damage. For small paddocks "G-lime" (Lafarge group) is a more expensive but perhaps a more convenient product to use. Liming is not necessary for all however, and for pastures on calcareous soils or other soils with a pH of more than 7 it should obviously not be used as serious mineral imbalances may ensue. A soil test and recommendation by a responsible body should always precede the application of lime. Over-liming can have adverse effects upon the availability of many other nutrients, for example magnesium, and may adversely affect the calcium: phosphorus ratio in the grazed herbage (Kellon, 1998).

So far as the other main nutrients are concerned soil testing for phosphorus (P), potassium (K) and magnesium (Mg) levels is fairly standard procedure and ADAS (1995) and MAFF (2000) both give reasonable advice for supplementation should levels fall. There is no virtue (and unnecessary expense) in maintaining high levels of these nutrients and high potassium levels may inhibit the uptake of magnesium. In paddocks where substantial "rough" dunging areas have developed, these may contain substantial concentrations of potash (ADAS, 1995). Sulphur (S) is a nutrient which has become deficient in some areas as a result of "clean air" policies and the reduction in industrial pollution in recent years. It is important in the synthesis of amino acids in the animal's body. In some cases, low sulphur levels have been shown substantially to reduce the response to other fertilisers and additional sulphur as a fertiliser material may be advisable, particularly on grass which is to be cut for hay or haylage (Dampney, 2000). Herbage analysis to determine the ratio of nitrogen: sulphur should be undertaken before any such supplementation takes place.

The greatest unknown is nitrogen (N). Grass responds well to nitrogen fertiliser when soil temperatures and moisture supply are adequate, however, supplies from soil reserves vary substantially and, of course, the presence of nitrogen fixing clovers or other legumes may preclude the need for any supplementation at all. ADAS (1995) recommend up to 44 kg N / ha on horse pastures three to four weeks before turnout and subsequent applications as necessary. However, this is one area where experience is important, and if adequate growth occurs without extra N as fertiliser, then it may be sensible to omit it altogether. Soil samples are not a particularly useful guide since levels of available N (as nitrate) vary substantially according to conditions affecting the mineralisation of N from soil organic matter.

Farmyard manure, which has been composted and well rotted, if available, is an excellent traditional fertiliser for grassland and contains all the major nutrients (including sulphur) required, although in quite variable amounts. Stable manure is not the ideal fertiliser for grassland to be grazed by horses since it will almost certainly contain eggs or larval stages of equine parasites. If it is used then it should be applied to grassland which is to be laid up for hay or haylage in the following year (but even used in this way there may still be some risk of infection). The designation of large areas of the United Kingdom as "Nitrate Vulnerable Zones" took place in 2001 and all occupiers of agricultural land should familiarise themselves with the legislation and relevant codes of practice before applying fertilisers to grassland (DEFRA, 2002).

Sward species composition is an unknown factor for many horse keepers (Judd *et al.*, 2001) although most know the value of perennial ryegrass (*Lolium perenne*). Less productive species such as *Agrostis stolonifera* (creeping bent), *Festuca rubra* (creeping red fescue) and *Poa pratensis* (smooth stalked meadow grass or "Kentucky bluegrass") can also be valuable because of their hard wearing properties and their ability to form a thatch over the soil. This helps to prevent poaching damage in adverse conditions. In fact, permanent pasture is in many ways more suitable for horse grazing than ley grass (sown grassland, previously probably in an arable rotation). Gibson (1998) reported a study into the effects of horse grazing on species-rich grassland and concluded that "light or moderate" grazing by horses was not detrimental to sward species composition but that heavy grazing by horses and other species was detrimental. Rogalski *et al.* (1999) found that horse grazing over a period of eight years reduced the proportion of *L. perenne* in a sward but increased the proportion of *Dactylis glomerata* and *Festuca pratensis*. ADAS (1995) recommend the following for newly sown horse paddocks: diploid perennial ryegrass (*L. perenne*), creeping red fescue (*F. rubra*), tall fescue (*Festuca aruninacea*), smooth stalked meadow grass (*Poa pratensis*), Timothy (*Phleum pratense*) and wild white clover (*Trifolium repens*). Others also recommend crested dogstail (*Cynosurus cristatus*), which although not particularly productive, is very palatable to horses. Longland and Cairns (2000) discussed the work of Chatterton *et al.* (1989) and others and offered suggestions about how the choice of species in sown mixtures could minimise the incidence of laminitis in susceptible animals. Cocksfoot (*D. glomerata*), Timothy (*P. pratense*) and white clover (*T. repens*) contained lower levels of fructans (key plant chemicals thought to cause the onset of laminitis) in cool conditions than, for example, perennial ryegrass (*L. perenne*) and meadow fescue (*F. pratensis*). It should be possible therefore to design grass / legume mixtures incorporating "low fructan" producing species in order to minimise this condition in susceptible stock. Such research is currently under way using Dartmoor ponies, at the Royal Agricultural College (Dr H Moreton, personal communication).

Broadleaf species in horse pastures include white clover which, although being useful as a source of fertility as a result of its nitrogen-fixing properties, can also become very dominant in closely grazed areas. However, in their eight-year study in Poland, Rogalski *et al.* (1999) showed a reduction in *T. repens* following prolonged horse grazing. McGorum and Anderson (2002) suggested that excessive ingestion of white clover may be implicated in the onset of grass sickness. Longland and Cairns (2000) emphasised the value of legumes in relation to a reduced potential for laminitis. Younie (2003) refers to the qualities of herbs such as chicory, ribwort, sheep's parsley and burnet which are sometimes added to sown horse pastures to provide variety and to enhance mineral intake. Frape (1998) also suggests grass / herb mixes for sowing in strips in horse paddocks. Other broadleaf species (some of which are potentially poisonous) which frequently occur in horse pastures are itemised and well illustrated by the British Horse Society (2000). DEFRA (2003) also designate the following as "injurious weeds" under the Weeds Act (1959); spear thistle (*Cirsium vulgare*) creeping or field thistle (*Cirsium arvense*) curled dock (*Rumex crispus*) broad leaved dock (*Rumex obtusifolius*) and common ragwort (*Senecio jacobaea*) These are all common in horse paddocks. Advice on the effective control of all of these weeds is widely available and the main points summarised by DEFRA (2003).

Of most concern at the present time is ragwort (*S. jacobaea*) (Boatman *et al.*, 2000). This plant is probably the cause of more illness and death among farm animals and horses than all other poisonous plants occurring in the UK put together (Gill and Vear, 1958). The British Horse Society (2003) referred to about 500 equine deaths confirmed in 2001 and a projected 1000 in 2002 as a direct result of ragwort poisoning. Ragwort is particularly damaging and palatable to animals after it has been cut and wilted or after spraying with an herbicide. The poison (pyrrolizidine alkaloid) affects the liver and the effects are cumulative. Ragwort control is the responsibility of the occupiers of land but this obligation is widely ignored, regrettably even by many local authorities. Sheep grazing often proves beneficial when ragwort plants are in the rosette stage but cannot be generally recommended since they may eventually succumb to the liver toxin. However in other parts of the world (New Zealand, for example) this method is still popular (Betteridge *et al.*, 2000). Other methods of control are effective and well understood and include digging or pulling by hand or pulling by machine (for example, by using the Alvan Blanch "Eco-puller"). Cutting is not recommended since compensatory growth may occur subsequently, creating an even more vigorous plant in the short term (Meijden *et al.*, 2000). It is important to remove and (preferably) to burn all the dead plants after pulling and to avoid allowing dead plants to be conserved as hay or haylage. Chemical control is also effective and Dixon and Clay (2001) describe the effectiveness of a range of herbicide applications including 2,4-D, clopyralid and glyphosate applied with a weed wiper. Spot treatment too with citronella oil was very effective. DEFRA (2003) also give a range of effective methods for control. It is commonplace for landlords letting land for horse grazing to insert a clause in the agreement requiring the tenant to control ragwort (and, it might be surmised, also in the tenant's interests). The most important point in control of course is not to allow the plants to set seed since each can produce up to 150,000 viable seeds.

There should therefore be no difficulty in controlling this very damaging weed and it is only indifference on the part of many landowners, occupiers and local authorities which has created the present epidemic. New regulations currently being laid before Parliament are intended to strengthen the provisions of the Weeds Act (1959) by the introduction of a statutory code of practice. It will, however, still require a concerted effort on the part of many to overcome one of the greatest indigenous threats to the health and well being of both farm and equine livestock.

Fungi may also be responsible for poisoning problems with horses kept at grass. Fescue toxicosis and ryegrass staggers are both due to ergot-like toxins produced by endophytic fungi which exist in the grass plants (Frape, 1998). Endophyte enhancement has been used by plant breeders as a technique to overcome biotic or abiotic stress in grasses, particularly the ryegrasses and tall fescue. Endophyte poisoning is more common in the USA and New Zealand than the UK.

Effective sward management for horse keepers can be problematic since many have a fixed area on which to keep their stock with little opportunity to conserve any hay or haylage. This, coupled with the variable pattern of grass growth depicted in figure 1 (ADAS, 1995) creates a potential "feast or famine" scenario as the seasons progress. A system whereby part of the grass area could be conserved for hay or haylage (as would normally be the case on a livestock farm) and subsequently made available for an expanded grazing area would help to provide a more constant supply of nutrients. However the absence of this proviso in many cases creates the need for increased feed supplementation as the season progresses. Subdivision of the grazing area into two or more grazing paddocks however, does introduce the possibility of rotational grazing, coupled with rotational fertilising, topping, dung removal and many of the other desirable aspects of paddock maintenance and improvement. However, Judd *et al.* (2001) reported that such practice was rare. Strip grazing using electrified tape is also a possibility and, provided good access to water is maintained and a second fence used to allow recovery of the sward already grazed (as shown in Figure 15.2), the system can be successfully used for horses. There is little information about optimum sward grazing heights for equines in the scientific literature. Most information seems to be anecdotal, however the British Horse Society (2003) confidently recommend between 5 and 10 cm. Frape (1998) refers to the effects of stocking density on the growth of young horses. Those kept at high densities were forced to graze into the less digestible lower third of the grass canopy whilst those at lower densities were able to graze more leafy, palatable and nutritious part of the canopy above 8 cm. Parasitism and the condition which has become known as "horse sickness" is a strong possibility on land, which is heavily and continuously grazed by horses year after year, even with the systematic removal of dung. It is extremely desirable therefore to alternate horse grazing with other stock (sheep or cattle) or with conservation for hay or haylage, from one year to another. Frape (1998) advocates the complementary use of ruminant stock to clear up stemmy or senescing herbage, either alternating with or grazing together with equines.

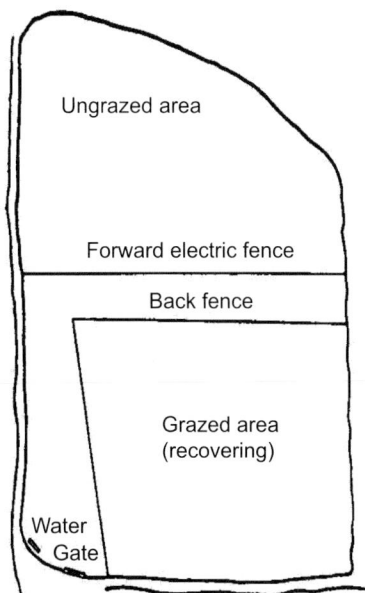

Figure 15.2.
Strip grazing (Finch *et al.*, 2002).

Regular paddock improvement activities (renovation, oversowing or complete reseeding) are engaged in by horse keepers but with variable success. Judd *et al.* (2001) reported that about 48% of the respondents to their survey regularly engaged in paddock "repair" or reseeding.

ADAS (1995) suggest that the optimum time for this in the UK is between July and early September. Important points are the suppression of the old sward with glyphosate spraying (or, of course, by ploughing) followed by liming and fertilising as suggested by MAFF (2000). Shallow seeding by broadcasting, drilling or raking-in is essential, followed by really good consolidation (Finch *et al.*, 2002). Control of slugs (*Deroceras* spp.) and frit fly (*Oscinella frit*) in the early stages of establishment are usually needed, followed by effective weed control particularly aimed at such weeds as chickweed (*Stellaria media*).

References

ADAS (1995). Grassland Management for Horses. *The ADAS Equine Team*. Agricultural Development and Advisory Service.

AFRC (1991). *Rothamsted Experimental Station: guide to the classical field experiments*. AFRC Institute of Arable Crops Research.

Archer, M. (1978). Studies on producing and maintaining balanced pastures for studs. *Equine Veterinary Journal,* 10: 54-59.

Betteridge, K., Costall, D. A., Hutching, S. M., Devantier, B. P. and Liu, Y. (2000) Ragwort (*Senecio jacobacea*) control by sheep in a hill country bull beef system. *Sheep Dairy News,* 17, No.1, 9-10.

Boatman, N. D., Clay, D. V., Goodman, A., Marrs, R. H., Marshall, E. J. P., Newman, J. R., Putwain, P. D., Pywell, R. F., (2000). Spread and control of common ragwort *Senecio jacobaea* in England. *Aspects of Applied Biology,* 58: 63-70.

British Horse Society (2000). BHS guide to grassland management for horse and pony owners. British Horse Society.

British Horse Society (2003). Some Equestrian Statistics [online]. Available from: http://www.bhs.org.uk/AccessLeaflets/Access24.htm [Date accessed 21.7.03].

Chatterton, N. J. Harrison, P. A. Bennett, J. H. Asay, K. H., (1989). Carbohydrate partitioning in 185 accessions of Gramineae grown under warm and cool temperatures. *Journal of Plant Physiology,* 134:169-179.

Dampney, P. M. R. (2000). Development and revision of national recommendations for the use of fertilisers and organic manures in England and Wales. *Proceedings - International Fertiliser Society,* No.457, 1-40.

DEFRA (2002). Guidelines for farmers in NVZs - England. Department for Environment, Food and Rural Affairs. PB5505. Revised July 2002.

DEFRA (2003). Guidance not on the methods that can be used to control harmful weeds [online]. Available from: http://www.defra.gov.uk/environ/weedsact/weedscontrol.htm [Date accessed 21.7.03]

Dixon, F. L. and Clay, D. V. (2001). Effect of synthetic and natural-product herbicides on *Senecio jacobaea* (common ragwort). *Proceedings -The BCPC Conference: Weeds 2001*, 721 - 726.

Finch , H. J. S., Samuel, A. M. and Lane, G. P. F. (2002). Lockhart and Wisemans's Crop Husbandry including Grassland 8th Ed. Woodhead Publishing Ltd.

Frape, D. (1998). Equine Nutrition and Feeding 2nd Ed. Blackwell Science, Oxford, UK

Gibson C. W. D. (1996). Effects of horse grazing on species-rich grasslands. *English Nature Research Reports*, 164. English Nature.

Gill, N.T. and Vear, K.C. (1958). Agricultural Botany. Duckworth.

Judd S. J., Seinet G. A. C., Boorman, A. V.,Morgan, J. E., and Whitaker, C. L. (2001). Horse pasture management study: report produced for the Countryside Agency by Writtle College. CAX 42. Countryside Agency.

Kellon, V. M. D. (1998). Equine Supplements and Nutraceuticals: a guide to peak health and performance. Breakthrough Publications.

Longland, A. C. and Cairns, A. J. (2000). Fructans and their implications in the aetiology of laminitis. 3rd *International conference on feeding horses*. Dodson and Horrell Ltd.

MAFF, (2000). Fertiliser Recommendations for Agricultural and Horticultural Crops (RB209). Ministry of Agriculture Fisheries and Food. The Stationery Office.

McGorum, B. C. and Anderson, R. A. (2002). Biomarkers of exposure to cyanogens in horses with grass sickness. *Veterinary Record*, 151:442-445.

Meijden, E., van der, Boer, N. J. de, Veen-Van Wijk C. A. M. van der (2000). Pattern of storage and regrowth in ragwort. *Evolutionary Ecology*, 14:439 - 455.

Odberg, F. O. and Francis-Smith, K. (1976). A study on eliminative and grazing behaviour - the use of the field by captive horses. *Equine Veterinary Journal*, 8: 147-149.

Rogalski, M.; Wieczorek, A.; Kardyska, S. (1999). Grazing animals as a factor regulating botanical composition of sward. *Folia Universitatis Agriculturae Stetinensis, Agricultura*, 75:267-270.

Thomas C., Reeve, A. and Fisher, G.E.J. (1991). Milk from Grass 2nd Ed. The British Grassland Society.

Younie, D. (2003). Natural Health. *Organic Farming*, 78: 28-29.

16

The challenge of attracting and retaining sponsorship in eventing

L Bowden and R Walsh
Royal Agricultural College, Cirencester, Gloucestershire, GL7 6JS, UK

Abstract

Eventing is a very expensive sport to organise and run and relies heavily upon sponsorship funding to enable it to remain financially viable. This research is concerned with the challenges that events face in attracting and retaining sponsorship and what can be done to address these problems, resulting in a clearer understanding of what makes a successful sponsorship. Fourteen interviews were conducted with both event organisers and sponsoring companies, as well four people who are considered to be knowledgeable about sponsorship within the equine industry, to ensure that a balanced view of the issues was gained. The research found that the industry suffers from a culture of patronage with some companies sponsoring eventing for altruistic reasons. They see sponsorship as a way of raising public awareness and of building goodwill with their customers by being seen to put something back into a sport that their customers care about. Sponsors commented that sponsorship impresses people and gives them an idea about the values of the company and where it is positioned within the industry. They saw the sponsored event as an opportunity for them to entertain clients, customers, press and other business professionals, often in very beautiful surroundings. As eventing attracts more AB and C1 competitors and spectators than any other sport, it can offer sponsors an opportunity for highly targeted marketing to a sector of the population who are acknowledged to be difficult to reach via traditional advertising. Sponsors are interested in getting value for money from their sponsorship, and they expect to be dealt with in a professional manner. Event organisers need to demonstrate clearly how the sponsorship will benefit the company and ensure that they meet all of the sponsors' business objectives. Sponsors expect to have name and brand awareness delivered to their target audience at the event, thus organisers must ensure that this happens by providing good branding opportunities. In order to gain the maximum benefit from the sponsorship, the sponsoring company needs to promote and publicise the sponsorship as much as possible by integrating it across their whole marketing strategy. This can be a great advantage to the event organiser as it raises the profile of the event. Events at the top end of the sport were found to have good sponsorship deals in place and the organisers said that they did not need sponsorship to make their events viable. However, the smaller events at grassroots level are only viable if sufficient sponsorship can be found. One interviewee suggested that these small events need to target smaller local companies who could benefit from raising their public profile in the local area. The event organisers interviewed feel undervalued by British Eventing, the sport's governing body, and think that British Eventing could do more to support event organisers in attracting and retaining sponsorship. This is particularly crucial at grassroots level, because without sponsorship these events are not financially viable.

The sport of eventing

Introduction

The sport of eventing (also called Horse Trials) is the ultimate test of all round horsemanship, where horse and rider must compete in all three disciplines of the sport; dressage, show jumping and cross-country. Historically, eventing is one of Britain's most successful sports in terms of reputation and the amount of medals won at major Championships – over 200 in 88 years of competition (British Eventing, 2002). Britain's enviable reputation as the mecca for the sport, both for hosting the best events and producing and training some of the best horses and riders in the world, has led to an enormous influx of foreign riders. It is estimated that 88% of all International event riders that leave their home country to live, train and compete abroad choose the UK (Ede, personal communication). The governing body of the sport, British Eventing, controls and regulates around 150 events annually. Of these, 10 are International three-day events, the most hosted by any one country in the world. Britain's flagship event, Badminton International Three-day Event, is regarded world-wide as 'the one the riders want to win'. It attracts an audience of over a quarter of a million over four days, with an estimated 100,000 spectators on cross-country day alone, making it one of the biggest annual sporting events in Europe. The sport is attractive for spectators as it is exciting, glamorous and, more often than not, set in the beautiful surroundings of Britain's most famous stately homes and parks. It is also one of the few sports where men and women compete on equal terms and genuine amateurs compete against World and Olympic Champions. British Eventing currently has 8800 members, 5000 of whom compete on a regular basis, and membership is increasing by around 5% per year (British Eventing, 2001). Eventing enjoys a clean

L. Bowden and R. Walsh

and sporting reputation, with relatively few scandals to tarnish its image, the biggest threat being the unavoidable risk of injury or death to horse or rider. However, overall, serious injuries and deaths to horses and riders are rare. The sport enjoys an unusually high demographic profile of AB spectators, riders and owners. Indeed, eventing attracts the highest number of AB adult participants of any sport at 63%, with the remainder falling into the C1 category. Of the spectators 52% are AB and 28% C1's (British Eventing, 2002).

The importance of sponsorship

Eventing is a very expensive sport to organise and run and events are coming under increasing financial pressure as costs continue to rise. According to British Eventing's Sport and Technical Manager, Tim Hadaway (2002), this is due to a number of factors:

1. Increased pressure to meet British Eventing guidelines and standards
2. The rising cost of medical and veterinary cover
3. Increasing standards in professional course design and construction
4. Increasingly competitive sponsorship marketplace

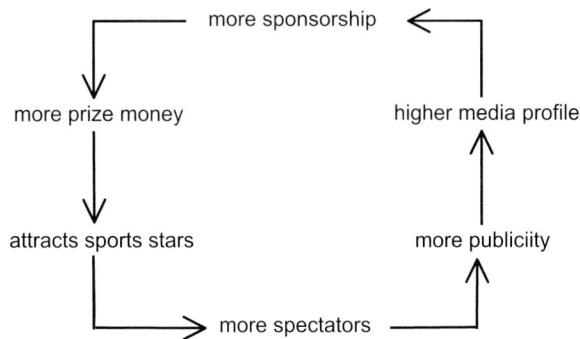

Figure 16.1. The virtuous cycle for successful events.

The importance of sponsorship as a revenue stream for events must not be underestimated. British Eventing recently stated that 'event financial viability is now the single most important issue facing the sport' (Hadaway, 2002). Sponsorship in eventing is currently stuck in a vicious circle where a lack of publicity and media profile means that there is a lack of interest from potential corporate sponsors. The sport needs to become more proactive in trying to break out of the vicious cycle and into the virtuous cycle where success breeds success. If the sport can attract more sponsorship funding, then this will result in more income for organisers and more prize money for the competitors. An increase in prize money will attract the sports' stars and result in high-class competitions which will result in greater spectator attendance figures and a greater demand for publicity. This in turn will increase the media profile of eventing and will generate more interest from potential sponsoring companies, thus increasing sponsorship funding in eventing. The virtuous cycle represents the interconnected factors necessary for a successful event and is illustrated in Figure 16.1.

The sport sponsorship industry

A definition of sponsorship

Traditionally, sponsorship deals belonged in the 'patronage' category and occurred when an outside body supported an event, sport or competitor without expecting any advantages or rewards for their money or resources. However, in today's highly commercial world, sponsorship has become redefined with the emphasis now firmly placed on achieving a commercial advantage and return on investment. Sleight (1989) confirms this by defining that "Sponsorship is a *business relationship* between a provider of funds, resources or services and an individual, event or organisation which offers in return some rights and association that may be used for *commercial advantage*". Although traditional 'patronage' type deals do still occur, more commonly in the arts or community related sectors, they are rare and are particularly unusual in the highly commercial sport sponsorship market.

An overview of the UK sport sponsorship market

There are three main categories of sponsorship in the UK marketplace, sport sponsorship, sponsorship of the arts and broadcast sponsorship. Of these sectors, sport sponsorship is the biggest and most mature segment of the market with companies choosing sport most frequently and spending to a higher value than in other sponsorship sectors. Sport has always been a natural arena for attracting sponsorship because of its many natural attributes:

- It has international appeal, therefore the ability to attract mass audiences
- It has the ability to break down national, cultural, geographical, class and language barriers
- It inspires high emotions and public heroes, thus making it newsworthy
- It has a rich variety of values and symbols
- It has two potential audiences, the participants and the spectators

Recent figures from the Key Note Report (Barker, 2001) show that the value of the UK professional sport sponsorship market in 2000 stood at £400 million, more than twice of what is spent annually on the arts (£150 million) and four times greater than the spend on broadcast sponsorship (£100 million). Estimates from Key Note (Barker, 2001) and Mintel (2000) predict that the market will be worth around £450 million in 2002 and around £500 million by 2005.

Sponsorship from a corporate perspective

In order to fully understand how to attract and successfully manage sponsorship programmes, it becomes necessary to comprehend sponsorship from a corporate perspective. Copeland, Frisby and McCarville (1996) note that "sport organisers may lack insight regarding the ways in which corporations select and evaluate sport sponsorship. Such information would enhance their understanding of the exchange process while increasing their ability to negotiate successful sponsorship initiatives."

Corporate sponsorship objectives

There are many reasons for companies to use sponsorship of sport as part of their promotional mix and overall marketing strategy. Meenaghan (1991) and Mintel (2000) found that the two most commonly cited reasons for sponsoring sport were to increase public awareness and to enhance a company or brand image. These, and other important corporate objectives for entering into a sponsorship agreement are discussed below.

- *To increase public awareness of a company or brand* – This is one of the main reasons for companies to use sponsorship as part of their marketing strategy. Sponsorship is particularly effective in increasing awareness because it is a more subtle form of advertising and works on a subliminal level. Spectators or viewers of an event are captive audiences where exposure to a sponsor's message is unavoidable because it is visible during viewing or attendance times.
- *To communicate company or brand values* – Sponsorship influences public opinion by endowing companies with positive images through an association with the things that consumers care about, thus allowing an affinity with the sponsors ethics, morals and concerns. Sponsorship is perceived by consumers to be good corporate citizenship where companies are seen to be 'putting something back' into society. This is becoming increasingly important with the rise of the ethical consumer (Marketlink Research, 1997).
- *To enhance or reposition company or brand image* – Reputation and image are the most important intangible assets that a firm possesses and sponsorship can be a useful tool to enhance or

reposition the company image. According to Ferrand and Pages (1996) the dichotomy of common and unique characteristics between a sponsoring company and an event offers companies two strategies. They can enhance their company's image by emphasising the common characteristics between the company's image and the event's image; alternatively, they could reposition their image by using a displacement strategy. By encouraging the unique image of the event to be transferred to the sponsor, the company's image can be effectively repositioned. Meenaghan and Shipley (1999) state that the sponsor and the sponsored event become involved in a symbiotic relationship that allows the events inherent values to be passed onto the sponsor as the audience are constantly exposed to logos and names throughout the event.

- *To provide a lifestyle association* – Sponsorship gives organisations the chance to align their image or brand with a particular lifestyle association. Every sport offers companies the chance to align themselves with a unique image combining many basic values, for example, glamour, fitness, excitement, skill, strength etc. Sport has particular appeal because it tends to be perceived as young, dynamic and evoke high emotions amongst spectators and fans.

- *To increase media coverage* – This is also an important reason for companies to utilise sponsorship. Television exposure tends to be regarded as the most important media, especially for mass-market consumer goods, and exposure to the sponsoring company's name and logo is unavoidable during the viewing of a sporting event.

- *To access niche markets* – Sponsorship can be used to effectively target niche markets because it is highly flexible and can be tailored to meet the needs of the audience. Certain markets, such as young males or those in the AB socio-economic bracket, are notoriously elusive to traditional advertising and respond better to sponsorship as a form of communication (Marketlink Research, 1997). Companies are able to directly target these niche markets by sponsoring the sports that appeal to them.

- *To increase sales and profit* – To have a positive impact on the bottom line is the underlying reason for any company to get involved in a sponsorship agreement. Sponsorship is seen as a business decision in today's competitive marketplace; therefore it must have an impact on sales and profit for a company to continue utilising it (Market Assessment International, 2000).

- *To improve customer loyalty* – Customer loyalty occurs when consumers believe that the value for money and quality of the sponsoring company's product is superior to that of its rivals. Customer loyalty may be achieved through sponsorship as consumers admit to feeling benevolent towards the sponsoring company for enabling the sporting event to take place through financial support (Marketlink Research, 1997).

- *To provide opportunities for corporate hospitality* – Sport sponsorship offers exciting and unique opportunities to entertain important clients, potential clients, buyers, suppliers and opinion formers.
- *To develop a relationship with consumers* – Sponsorship is thought to be particularly effective for building a relationship with consumers because it allows them to form an emotional connection with the sponsors, thus encouraging brand loyalty and improves corporate image.

Research methodology

This research comes from a positivist philosophy because the researcher remained independent and objective towards the study and was not concerned with people's interpretations of their actions, but with the actual relationships between event organisers and sponsors. The approach to the research was of a qualitative nature because it was felt that this would lead to a deeper understanding of the subject by asking why eventing struggles to gain sponsorship, and how this situation can be improved. By developing some generalisations and ideas about the relationships between event organisers and sponsors, it became possible to suggest practical solutions and ideas about improving them. Because this research adopted an interviewing strategy based on collecting and analysing people's experiences and thoughts on the sponsorship issue, it was considered unsuitable to attempt to quantify the data statistically. Throughout the research process of conducting interviews and reviewing the existing literature, the theory that the success of a sponsorship lies with the critical relationship between the event and the sponsor was developed inductively. Consequently, an interviewing strategy, involving both event organisers and sponsoring companies, was developed in order to gain a balanced and in-depth view of this critical relationship. The use of in-depth interviews was selected in order to generate rich and reliable primary data through relatively informal and conversational interaction between the interviewer and interviewee. In order to gain a perspective on the sponsorship relationship from both viewpoints, five event organisers seeking sponsorship and five companies who sponsor equestrian sports were interviewed. The remaining four interviewees were chosen because of their authority on the subject; for example, an enquiry into the sponsorship of eventing would have been incomplete without an interview with British Eventing and their marketing and sponsorship consultants, the Grand Plan Consultancy (GPC). Interviews with two comparable equestrian sports, dressage and racing, were undertaken to try and gain an insight into how other disciplines have dealt with the issues. Face-to-face interviews with sponsors and organisers who represented both ends of the spectrum, covering grassroots organisers

THE LIBRARY
BISHOP BURTON COLLEGE

and sponsors through to large corporate sponsors and the sports' showcase events. It must be recognised that any generalisations made about attracting and retaining sponsorship are based upon interviews with five organisers, five sponsors and four other people who are involved with eventing or sponsorship of another equestrian sport.

Discussion

Why companies sponsor

Literature has suggested that sponsorship is now accepted as a valuable addition to the marketing mix, and is used to achieve a number of corporate objectives. Meenaghan (1991) and Mintel (2000) found that increasing public awareness and enhancing corporate image were the two most common reasons for a company to utilise sponsorship in their marketing strategies. Other objectives found in the literature were communicating company values, providing lifestyle associations, increasing media coverage, accessing niche markets, differentiating companies from competitors, gaining a competitive advantage, increasing sales and profit, improving customer loyalty, opportunities for entertaining and boosting staff morale. Most of these objectives were mentioned by the sponsoring companies interviewed, all of whom mentioned raising public awareness of their brand or company as one of the most important reasons for sponsoring. Image enhancement can be achieved by an association with a complementary event that fits the company's desired reputation. For example, sponsorship can be used to impress the target market by the company demonstrating a willingness to put money into a sport that their clients love. The sponsoring companies interviewed commented that sponsorship impresses people and gives them an idea about the values of the company and where it is positioned within the industry. They also commented that, by being seen to put money back into the industry where they make their profit, it is a very good way to build goodwill with their consumers. This idea is supported by Marketlink Research (1997) who believe that the rise of the ethical consumer has forced companies into behaving morally by showing that they are prepared to invest money and behave less selfishly. The opportunity to entertain key clients, customers, press and other business professionals was also rated highly as a sponsorship objective. The sponsoring companies use customer hospitality marquees at their sponsored events to encourage interaction with their target market, as well as entertaining key business associates and potential clients in interesting and prestigious surroundings. Sponsorship also offers the opportunity for highly targeted marketing to a captive audience, or to niche markets that are difficult to reach through conventional marketing techniques, by sponsoring the activity that they enjoy. This was felt to be a very important objective for several of the sponsors interviewed who said that they only sponsor activities that their target audience are directly involved in. Sponsorship allows interaction between sponsors and clients, which the sponsors felt to be important, by allowing opportunities for demonstrations, road shows and merchandising. One of the sponsors felt that sponsorship should be used as a 'shop front' with point-of-sale opportunities, product placement stands and live demonstrations allowing a more interactive method of communicating with their clients. It is widely believed that this interaction with consumers allows them to build a relationship with the sponsoring company, thus improving loyalty and corporate image (Marketlink Research, 1997).

What sponsors expect

Sponsorship today requires a commercial and professional approach and sponsoring companies expect to get a return on their investment and to have their sponsorship objectives met. British Eventing (personal communication) advise organisers to go into sponsorship deals knowing what the sponsors want, and confident that this can be achieved. They feel that event organisers must try to meet the sponsor's objectives as closely as possible. The sponsoring companies interviewed supported this view, all of them review their sponsorship activity on a regular basis to ensure that it is meeting their requirements and adding value. The GPC (personal communication) say it is important that event organisers remember that 'a good sponsorship is about creating a win: win situation – not about a company just "donating" money to an event.' Other interviewees agreed and pointed out that events must have 'something to sell', in other words sponsorship proposals need to be realistic in offering the potential sponsors some real benefits and demonstrate clearly how the sponsorship will be of value to them. Most interviewees felt that offering sponsors value for money was very important in ensuring that eventing is able to continue to attract decent sponsorship now and in the future. The GPC note that everyone involved in sponsorship must ensure that sponsors get value for money, offered in professional sponsorship packages, as this is what sponsors expect above all else. Eventing has to compete not only with other marketing options, but with other sports as well for sponsorship funding. Thus, it must offer value for money and demonstrate it in real terms. The interviewees feel that all sponsors expect to have name and brand awareness delivered to their target audience, therefore, it is essential that events are able to deliver the promised audience through the events viewers, participants, spectators and readers. Event organisers must be able to provide their sponsors with good branding opportunities and as much media coverage as possible so that the sponsors' objectives of raising awareness of their brand can be met. There is some concern about sponsors gaining value for money and a return on investment at the lower levels of the sport. Both grassroots organisers and organisers of the larger events agreed that it is becoming increasingly difficult to attract sponsorship to the grassroots end of the sport because there is not really anything to offer to sponsors in terms of spectator numbers and media coverage, thus it is difficult to make proposals unique and interesting.

The need for sponsorship in eventing

The importance of sponsorship for eventing to remain financially viable was a constant theme throughout the interviewing process. The GPC (personal communication) point out that good sponsorship is beneficial in more ways than just money. They feel that an association with an organisation with complementary brand values and a professional marketing department can be advantageous to an event and the reputation of the sport as a whole. This is supported by Geldard and Sinclair (1996) who noted that the sponsor has an obligation to promote the interests of the sponsorship wherever possible. Hastings (1984) also feels that the company must promote the event, and therefore its sponsorship, in order to meet their sponsorship objectives. British Eventing (personal communication) believe that sponsorship is very important, especially in terms of prize money, facilities and glamour, but also comment that there is already a reasonable amount of sponsorship in eventing. The unfortunate spate of rider fatalities during the 1999 season has justifiably resulted in increased safety standards, and therefore, costs. Vets and doctors, who were traditionally paid with a bottle of Whisky, now charge between £250 and £350 per day and paramedics require four-wheel drive vehicles, and in some cases helicopters which cost between £1000 and £2000 per day (British Eventing, personal communication). The most expensive item for events is the construction, upgrading and maintenance of the cross-country courses, which are having to be built to an increasingly higher standard due to safety concerns, thereby increasing costs further still. One of the grassroots organisers paid approximately £20,000 for a novice course to be built. Courses for intermediate, advanced and three-day events cost much more. There is a big difference between the events at grassroots level and the top one and three-day events. Most events, with the exception of the top three-day events and some of the more prestigious one-day events, do not charge spectators gate money. This is often because there are not enough spectators to make it worthwhile, and advertising to attract more is very expensive. A lot of events are also held on weekdays and most of the spectators are event horse owners, who already have car passes as part of their membership fees. The lack of spectators makes it difficult for these events to attract trade stands, thus they gain no income through the rental of pitches. British Eventing (personal communication) estimates that while Badminton has a turnover of £4 million, grassroots events will not make much more than £5,000. However, this may be an overestimate and all the organisers of the smaller events interviewed say they are lucky to break-even and that their events would not be financially viable without sponsorship. One of the organisers interviewed believes that the sport as a whole is not in a position to run without sponsorship and that the escalating costs of running horse trials is resulting in an even greater need for sponsorship, particularly at grassroots level. Another organiser summed up the situation by stating 'I think our figures show that it is impossible to make money without sponsorship, especially if you have to pay rent and labour costs.' (Batten, 1999). Although both organisers interviewed who run international three-day events, said their events could run without sponsorship, they said it would be to the detriment of British Eventing. Currently, these two major three-day events represent over 60% of British Eventing's annual turnover (Durrant, 2002), therefore, if they lost their sponsors British Eventing would lose a substantial portion of their income to distribute to other events and the sport as a whole would suffer.

The culture of sponsorship in eventing

According to literature, sponsorship has moved away from traditional patronage, whereby companies altruistically donate money without expecting any return on investment, and is now a business partnership between the event and the sponsor. Sponsors expect a professional approach to sponsorship whereby they receive value for money and commercial benefits in return for a sponsorship fee. According to Bowdin *et al.* (2001), sponsorship is used by companies for purely commercial and promotional reasons and has nothing to do with philanthropy or donations. However, it appears that within the eventing sponsorship marketplace, a culture of patronage still exists. According to the interviewees, most sponsorship deals are gained through word of mouth, chance, an association with eventing or a personal interest, rather than structured and cohesive sponsorship plans. Encouragingly though, there is a growing realisation of the need to become more commercial in order for the sport to continue to move forward. The GPC (personal communication) believe that the sport needs to move away from the culture of patronage and develop a more professional approach to sponsorship, by developing clear brand values that sponsors want to be associated with.

What eventing can offer

In order for any sport to gain sponsorship in today's competitive and commercial marketplace, it must have something to offer to potential sponsoring companies. There was some concern amongst the interviewees that eventing is not a particularly sponsorable sport at the lower levels where there is not much to offer to sponsors in the way of media coverage and spectator numbers. One organiser of an international event feels that it is 'a hopeless and silly task to think that small events will have decent sponsorships', but also commented that there is potential for smaller events to gain sponsorship at a lower level from local companies 'if the value and price is right.' Other interviewees felt that there is plenty of scope for sponsorship at grassroots level, but feel that it needs to be done at a local level with a greater number of smaller sponsors, rather than expecting one big sponsorship deal. One interviewee suggested that grassroots organisers approach 'local people who want to be a big fish in a small local pond'. This view is supported by research from RQA Ltd (1999) who suggest that there is a market for low cost, single event sponsorship at grassroots level. Overall however, it is felt that eventing has attractive brand values to offer to sponsors and that there is the potential to attract more sponsorship. The interviewees felt that the sports' main qualities are excitement, fast-paced, glamorous, equality between men and women, partnership between horse and rider and an element of danger. Two of the sponsors interviewed said that eventing has synergy with the qualities that their companies wish to portray. For example, one of them feels that the sport is skilful, athletic, dynamic, exciting and is set in venues that compliment her company's image as a slick organisation and the market leader. The other sponsor comments that eventing fits perfectly with their brand values of guts, authenticity, adventure and supremacy and that the sports' long-standing, well-respected and traditionally British heritage also fits with the company's core values. Another advantage for eventing seems to be the demographic profile of the audience and the spectators and the fact that it has royal connections. One interviewee commented that eventing has a 'huge advantage' because of the royal connections and that Princess Anne's participation in the sport in the seventies meant it was widely marketed to the general public. Eventing offers highly targeted marketing to the niche market of AB professionals who are notoriously difficult to access through traditional advertising techniques (Barker, 2001 and Marketlink Research). The GPC (personal communication) are currently approaching potential sponsors who want access to the affluent and educated consumers who make up the eventing audience. For example, they are approaching companies in the banking, IT, motoring, luxury goods and property industries. However, although the demographic profile and royal patronage can be advantageous, there is concern that the sport has an elitist and inaccessible reputation, which can also deter potential sponsors, the media and the general public. British Eventing admit that whilst they are trying to push the royal patronage and demographic profile of the sport to suitable sponsors, they realise that there is a potential conflict because they are simultaneously trying to get rid of their elitist reputation (personal communication). Eventing can also offer sponsors something special in terms of corporate hospitality opportunities. Many of the events are held in some of the most beautiful private parks and stately homes in Britain. British Eventing (personal communication) feel that this is one area that eventing needs to capitalise on. The GPC point out that the sport can offer expansive corporate hospitality opportunities and that 'everything from a picnic on the cross-country course to a full blown four course lunch with champagne reception or party in a large country houses is possible' (personal communication). Indeed, all of the sponsoring companies interviewed said that the opportunity to entertain guests in interesting and spectacular settings was amongst their key objectives when entering into sponsorship agreements.

The relationship between sponsors and event organisers

Ultimately the success of a sponsorship lies with the critical relationship between the organisers of the event and the sponsors. Geldard and Sinclair (1996) point out that both parties have a mutual obligation towards each other to ensure that the sponsorship is a success. The organisers that were interviewed appear to realise this, although most of them knew of sponsorship deals that had been lost simply through bad people management. There was some concern that organisers, particularly at grassroots level, were not sufficiently aware of the need for professional, business relationships with sponsors. One sponsor commented that she had turned down a sponsorship proposal because the organiser did not appear to be interested in their business objectives and was very arrogant. She believes that this can be a problem in eventing. Overall however, the organisers that were interviewed realised that it was their responsibility to make the sponsorship successful and believed that regular communication with the sponsors was one of the keys to success. They felt that having a good, close, business relationship with their sponsors is vital. One organiser said that problems arise when; sponsors are not kept informed, are asked for more money than was originally agreed, events do not deliver on their promises or exaggerate what they can offer. British Eventing (personal communication) advise organisers to enter into sponsorship agreements knowing exactly what the sponsors require and feeling confident that they can meet their business objectives. Geldard and Sinclair (1996) also state that events must undersell and over-deliver on their promises to sponsors to ensure success. It is felt that good preparation is essential when dealing with sponsors so that organisers can understand the company's motives for sponsoring, thus be better able to help them meet their objectives (Crompton, 1993). Researching the sponsoring company will allow organisers to work out ways in which to benefit the sponsor, for example, they need to be 'prepared to bend over backwards to sell the opportunity', as one organiser commented. Research from RQA Ltd (1999) demonstrates that sponsorship proposals do not clearly identify benefits and returns on investment for the sponsoring companies or specify the demographics of the audience. It is believed that proposals need to be researched and then tailored to fit the individual company's objectives thus offering them real benefits and advantages. British Eventing (personal communication) say that 'happy sponsors will return' and another interviewee believes that 'a sponsor goes where he's solicited and stays where he's appreciated'. One organiser achieves this by having a dedicated and knowledgeable team of helpers at his event whose sole job is to attend to the sponsors. He says it is essential that sponsors get the recognition they deserve and are treated as important people at all times and that they are more likely to continue the sponsorship if they have an enjoyable time at the event. Another organiser commented that she takes all the responsibility away from her sponsors so that all they have to do on the day of the event is enjoy themselves. Projecting a professional image is paramount in order to impress sponsors and keep the business relationship healthy. For example, it is vital that organisers approach potential sponsors well in advance, before their marketing budgets are set, rather than leaving it to the last minute. Overall, the interviewees seemed to think that sponsorship in eventing is managed reasonably well, although there is always room for improvement, but that it also depends upon the individual organiser, as some are better than others.

Conclusions

British Eventing have recently announced that 'event financial viability is now the single most important issue facing the sport' (Hadaway, 2002). Thus, sponsorship is essential for the success of both individual events and British Eventing if the sport is to continue to expand and sustain new growth, especially in this increasingly unstable economic climate. Although the sports' flagship events are currently capable of running without sponsorship funding, British Eventing would lose a substantial portion of their income if these events were not sponsored. Sponsorship is even more important at the rapidly expanding lower levels of the sport where organisers insist that their events are not financially viable without sponsorship. Although literature suggests that sponsorship is no longer about companies altruistically donating money to an event, it would appear that a culture of patronage and donation still exists. Eventing appears to rely heavily upon this culture, particularly at grassroots level where events do not have a lot to offer to sponsors in terms of media coverage and attendance figures. However, there is potential for sponsorship at a local level and grassroots events are still able offer the sport's attractive brand values, access to affluent niche markets and corporate hospitality opportunities. Event organisers must realise that the traditional culture of patronage is becoming an increasingly commercial sponsorship environment, thus requiring a professional approach. Therefore, event organisers must not continue to rely upon patronage but seek to become more businesslike by being realistic about what they can offer and ensuring that they understand and are able to fully satisfy their sponsor's business objectives. Companies these days expect to get a return on investment and value for money, thus if events are unable to deliver these benefits then the sponsorship will be terminated.

Organisers must realise that they are ultimately responsible for the success of their sponsorship, which lies with the critical relationship between them and their sponsors. Organisers must appreciate that sponsorship is a business relationship between the two parties, and that they must ensure that they know, understand and deliver the sponsor's business objectives. It is suggested that organisers identify and approach companies who have synergy with eventing and who wish to access the sports affluent and educated audience. Research should be undertaken before approaching potential sponsors in order to understand their motives for sponsoring and what benefits they will require. This will ensure that sponsorship proposals are individually tailored and realistic, outlining specific benefits to the potential sponsor. Proposals should ideally be professional documents that are submitted in plenty of time, taking account of the budgetary cycles of companies, and presented on a face-to-face basis. Sponsors must be made to feel valued at all times and treated in a professional manner, they expect to be kept informed of all decisions, shown respect at all times and made to feel important by involving them in the overall event. Whilst it is understood that it can be difficult to attend to the sponsors at all times when the event is underway, particularly for grassroots events who have small teams and limited resources, having a dedicated team to attend to the sponsors is vital. If sponsors are given an enjoyable experience at the event, for example, guided course walks and social events with competitors in attendance, they are much more likely to sponsor the event again. Conflict between sponsoring companies must also be avoided if the sponsors are to remain satisfied, as no sponsor wants to be overshadowed by their rivals sponsoring at the same event.

British Eventing recently stated that organisers and their volunteers are the 'heart of the sport' and that they are 'truly indebted' to each one of them (Durrant, 2002). However, whilst it is acknowledged that this research investigated a very small sample of event organisers, it appears that organisers feel undervalued and unappreciated by British Eventing, especially at grassroots level. The organisers are mostly amateur volunteers with little or no relevant business experience who run events for altruistic reasons and to put something back into a sport that they love. Currently, the individual organisers are left to find their own sponsors, and whilst it is appreciated that it would be difficult for British Eventing to find sponsors for all events, it is logical to think that British Eventing should be more involved than they are. After all, British Eventing is the governing body of the sport and thus has a responsibility to be the central resource for event organisers. According to British Eventing's Sport and Technical Manager, the most significant area of growth in the sport has been at these lower levels, particularly at intro level where the number of starters has expanded by a staggering 160% (Hadaway, 2002). Whilst it is acknowledged that British Eventing must satisfy this demand and exploit this valuable income stream, it is suggested that they should be more proactive in supporting the organisers at this level to find vital sponsorship. It is appreciated that these grassroots events have little to offer to sponsors in terms of spectator numbers and media coverage, therefore it is not being suggested that these events will ever attract huge amounts of funding from large corporations. However, they do require more proactive and hands-on support to find local sponsors in order to ensure that they are able to become financially secure enough to continue to provide the grassroots events that are so necessary for the sport.

Eventing currently suffers from a low profile and a lack of media and television coverage that could discourage potential sponsors. Part of this problem stems from the fact that the sport has been poorly marketed in the past, it is difficult to understand and has an elitist reputation. Although television coverage is seen as an important objective for many potential sponsors, it is becoming increasingly more difficult for eventing to attract television because of the declining viewing figures. If the sport is to capture the imagination of the general public, gain more media attention and, therefore, more corporate interest and funding, then it must be marketed efficiently to other markets outside of the equine industry. However, it is worth noting that eventing is a minority sport that will never have the ability to attract the same amount of media coverage and corporate funding as some other mass-market sports, for example, football. On a more positive note, eventing has a number of attractive brand values, such as excitement, danger and a clean reputation. It can offer expansive corporate hospitality opportunities and it has an affluent and educated ABC1 audience which it can offer to potential sponsoring companies who wish to access niche markets. British Eventing seem to have recognised the need for more effective marketing in order to attract more sponsorship in the future, and have admitted that this was a weakness in the past. Part of the GPC's remit is to attract new sponsors by increasing the marketing of the sport and developing a brand awareness of the sport's attributes.

The recent threat of an economic downturn means that British Eventing and event organisers may have to strive harder than ever to attract and retain sponsorship funding. British Eventing need to ensure that they have a structured and cohesive plan in place to collaborate with event organisers and fully support them by ensuring that they are able to attract and retain sponsorship throughout a recession. Whilst some companies may feel that their sponsorship is a vital part of the marketing mix, on the whole a downturn in the economy will result in marketing budgets, and therefore sponsorship, being reduced. Consequently, demand for sponsorship will exceed the supply of sponsors. Thus, event organisers must realise that the retention of current sponsors is likely to become critical to their success, therefore, the ability to induce loyalty and maintain a healthy business relationship with sponsors is absolutely paramount.

References

Amis, J., Slack, T. and Berrett, T. (1999). Sports sponsorship as distinctive competence. *European journal of marketing.* 33(3/4): 250-272

Barker, L. (ed.) (2001). *Sports sponsorship – 2001 market report.* 2nd ed. Key Note Publications

Batten, S. (1999). Interview in, Mortimer, C. 1999. Counting the cost. *Eventing.* December: 15

Bowdin, G., *et al.* (2001). *Events management.* Butterworth-Heineman

British Eventing. (2002). *Home page.* [online]. Available from: http://www.britisheventing.com [Accessed: 12/5/02]

Copeland, R., Frisby, W. and McCarville, R. (1996). Understanding the sport sponsorship process from a corporate perspective. *Journal of sport management.* 10: 32-48

Crompton, J.L. (1993). Understanding a businesses organisation's approach to entering a sponsorship partnership. *Festival management and event tourism.* 1(3): 98-109

Durrant, P. (2002). *Chief executive's annual general meeting speech.* [online]. British Eventing. Available from: http://www.britisheventing.com/news.html [Accessed: 30/10/02]

Ferrand, A. and Pages, M. (1996). Image sponsoring: a methodology to match event and sponsor. *Journal of sport management.* 10: 278-291

Geldard and Sinclair. (1996). In, Bowdin, G. et al. 2001. *Events management.* Butterworth-Heineman

Hadaway, T. (2002). *Sport and technical manager's annual general meeting speech.* [online]. British Eventing. Available from: http://www.britisheventing.com/news.html [Accessed: 30/10/02]

Hastings, G.B. (1984). Sponsorship works differently from advertising. *International journal of advertising.* 3: 171-176

Market Assessment International. (2000). *Sponsorship 2000.* Market Assessment International

Marketlink Research (1997). *Sponsorship into the millennium: current trends and the future of sponsorship – a research report.* Marketlink Research

Meenagahn, T. (1991). The role of sponsorship in the marketing communications mix. *International journal of advertising.* 10: 35-47

Meenagahn, T. and Shipley, D. (1999) Media effect in commercial sponsorship. *European journal of marketing.* 33(3/4): 328-347

Mintel International Group Ltd. (2000). *Mintel Special Report: Sponsorship 2000.* Mintel International Group Ltd

RQA Sport and Leisure Consultants Ltd. (1999). *Business sponsorship in sport.* Sports Sponsorship Advisory Service

Sleight, S. (1989). *Sponsorship: what it is and how to use it.* McGraw-Hill Book Company. 4

Preweaned Thoroughbred (*Equus caballus*) foals exhibit fewer behavioural signs of distress if accompanied by a handler while separated from their dam during covering

S McGee and H V Smith

Department of Psychology, Aras an Phiarsaigh, Trinity College, Dublin 2, Ireland, mcgees@tcd.ie

Introduction

Thoroughbred mares (*Equus caballus*) are usually first teased 6 days post foaling and during this brief (often 3-6 minutes) process the mare's foal will be left alone in its stable. This will be repeated daily until the mare goes out of season. During this part of her cycle she may later be sent for covering by the resident stallion. On some farms the foal is alone in its stable during covering and this period can last for anything from 20 minutes to an hour, during which the foal typically exhibits signs of considerable distress. Though the foal can be heard whinnying for the mare, it has not previously been noted what behaviours the distressed foal exhibits during this separation. Though a recent study by Houpt (2002) looked at responses of 10 pony mares and their foals to brief weekly periods of separation at up to nine weeks of age, most research on separation distress in foals has previously been done with respect to the artificial weaning process, an event that most Thoroughbred foals experience by 6-7 months of age. In horses, stress has repeatedly been linked to the development of stereotypies, abnormal behaviours that often appear to lack an obvious function and which occupy a large part of the animal's time (Mason, 1991; Houpt, 1998). Kiley-Worthington (1997) notes that behavioural problems in horses can often be traced back to the time of weaning but it is possible that in Thoroughbreds these repeated separations in the first few weeks of life may also have a role to play. Between 2.5 and 5.5 per cent of the breed develop stereotypies (Vecchiotti and Galanti, 1986; McBane, 1994) though whether this is due to heredity factors or the less than ideal circumstances in which they live (Kiley-Worthington, 1997) is not yet clear. What is clear is that their monetary value and desirability as potential racehorses diminish if these behaviours develop. This study sought to determine if preweaned Thoroughbred foals would exhibit fewer behavioural signs of distress if accompanied by a handler while separated from their dam during covering.

Materials and methods

Fifty-seven preweaned Thoroughbred foals were assigned to either the accompanied (m=10; f=17; age range = 10-64 days; M=33 days) or isolated (m=15; f=15; age range = 9-97 days; M=35 days) group as they became available for observation. In the accompanied condition the observer entered the stable as the horsebox containing the mare was leaving the yard. The observer did not interact with the foal unless the animal initiated contact. Observations were recorded with pen, paper and stopwatch in order to minimise the amount of disturbance to the foal. In the isolated condition the observer viewed the foal from outside the stable, either by viewing through the window or though the bars on the door (while standing to the left of the door, out of view of the foal). Data were collected in the same manner as in the accompanied condition. In both cases data collection terminated as the horsebox returned. Behavioural signs of distress were recorded as vocalisations, time spent pacing or moving about (bouts of 3+ continuous steps), time spent pawing and the number of times the foal struck a solid object.

Results

The interval between vocalisations was significantly (z=3.88; p<.001) shorter for foals in the isolated group (mean = 14.8 sec; s.e.m=2.2 sec) than in the accompanied group (mean = 26.8 sec; s.e.m=2.9 sec). Foals in the accompanied group spent a smaller proportion of the time observed (13.7%; s.e.m=2.2%) engaging in pacing or moving about the stable than foals in the isolated group (20.7%; s.e.m=3.6%) though the difference was not significant. Only 12 of the 27 accompanied foals pawed (mean = 6.8 sec; s.e.m=2.4 sec) while 21 of the 30 isolated foals engaged in the behaviour (mean = 29.6 sec; s.e.m=9.9 sec); this was a significant difference (z = 2.50; p=.012). A total of 17 of the 30 isolated foals hit themselves against a door or wall at least once during the period of observation but none of the accompanied foals engaged in the behaviour ($X^2 = 19.18$; df=1; p<.001). Eight of these foals hurled themselves against the door between 11 and 67 times. One colt also showed signs of weaving. Foals that vocalised more often were more likely to move about (r_s=-.38; n=57; p<.01) and to strike a solid object (r_s=-.42; n=57; p=.001). Foals that spent more time moving around were more likely to strike solid objects (r_s=.43; n=57; p=.001). Foals that struck solid objects were more likely to paw (r_s=.37; n=57; p<.01).

Conclusions

Having a human handler stay, quietly, in the stable with a preweaned foal during the time its dam is absent for covering reduces the behavioural signs of distress exhibited by the animal and eliminates potentially injurious behaviours such as charging into a wall or door. Therefore, by taking the time to help calm a panicked foal rather than leave it to its own devices, the risk of injury and, potentially, a trigger for the development of abnormal behaviours, is reduced.

References

Houpt, K.A. (1998). *Domestic Animal Behaviour for Veterinarians and Animal Scientists.* (3rd Ed) Manson Publishing Ltd, London, UK.

Houpt, K.A. (2002). Formation and dissolution of the mare-foal bond. *Applied Animal Behaviour Science* 78: 319-328.

Kiley-Worthington, M. (1997). *Equine Welfare.* J A Allen, London, UK.

Mason, G.J. (1991). Stereotypies: a critical review. *Animal Behaviour* 41: 1015-1037.

McBane, S. (1994). *Behaviour Problems in Horses.* David & Charles, Devon, UK.

Vecchiotti, G.G. and Galanti, R. (1986). Evidence of heredity of cribbing, weaving and stall-walking in Thoroughbred horses. *Livestock Production Science* 14: 91-95.

Meal frequency and stereotypic behaviour in stabled horses

J J Cooper[1], S Johnson[1], N McCall[1] and N Davidson[2]
[1]Animal Behaviour, Cognition and Welfare Research Group, Department of Biological Sciences, University of Lincoln, Riseholme Park, Riseholme, Lincoln LN2 2LG UK, Email: jcooper@lincoln.ac.uk
[2]Equine Studies Group, Waltham Centre for Pet Nutrition, Waltham-on-the-Wolds, Leicestershire, UK

Introduction

Locomotor stereotypies such as weaving and nodding are commonly observed as pre-feeding activity in stabled horses (Cooper, McDonald and Mills, 2001), whilst oral activities such as crib-biting and wood-chewing may be associated with the post-feeding period (McGreevy and Nicol, 1998) particularly in horses fed a low-fibre concentrate (Nicol et al., 2002). In this study, we investigated the effect of increasing the number of meals of concentrate whilst maintaining the same daily concentrate intake, on the behaviour of stabled horses with particular reference to stereotypic activities such as weaving and crib-biting. The study was carried out on a working equestrian yard with stables for up to 50 horses.

Materials and methods

A pilot study was used to record incidence of stereotypic behaviour and to select subjects for the main study. In this, the behaviour of 30 warm-blooded horses was recorded during their morning (0830h) and afternoon (1630h) concentrate meals. Each meal consisted of 0.5kg of a commercial feed, mixed with 1kg of soaked sugar beet pulp and 0.5kg of chopped hay and straw mixed in a bucket. Horses were observed from an hour prior to concentrate presentation to half an hour following concentrate presentation for two successive days.

In the main study 9 horses were fed their normal ration of concentrate divided between two (2kg), four (1kg) or six (0.66kg) meals per day. For two meals per day treatment horses were fed concentrate at 10.30am and 16.30 pm; for four meals per day they were fed at 10.30am, 12.30pm, 14.30pm and 16.30pm; and for six meals per day, feeding time was at 10.30am, 11.30am, 12,30pm, 14.30pm, 15.30pm and 16.30pm. Horses were observed for six hours per day from 10.00am to 13.00pm and from 14.00pm to 17.00pm. The behaviour of the treatment horses was compared with that of 7 control horses, which received two meals per day throughout the trial.

Results

In the pilot study, there was a low incidence of stereotypic behaviour (5.6% of scans) in the population. These were more commonly observed in the afternoon (7.1%) than the morning observations (4.2%; $P < 0.05$). The higher incidence in the afternoon appeared to be related to the lower availability of high-fibre forage during the afternoon meal as in the morning horses received haylage at the same time as concentrate, whilst in the afternoon, haylage was not added until approximately half an hour following presentation of concentrate.

In the main study, as the number of meals increased, the treatment horses showed a decrease in oral stereotypies ($P < 0.01$), but an increase in weaving ($P < 0.05$) and nodding ($P < 0.01$) prior to feeding (Table 1). The control group increased weaving, nodding and oral stereotypies (all $P < 0.05$) as their yard-mates received more meals. Consequently there was an overall increase in incidence of stereotypy in both treatment and control horses with the increase in meal frequency.

Table 1.
The percentage of scans engaged in feeding activities and stereotypic activities by treatment and control horses, when treatment horses receive their ration as 2, 4 or 6 meals per day.

	Treatment Horses				Control Horses			
Active feed	2	4	6	F	2	4	6	F
Conc.	5.7	7.8	8.7	16.5**	4.6	4.0	6.0	2.82
Hay	13.6	18.0	25.4	15.3**	12.3	17.6	14.3	2.30
Ssy								
Weave	0.27	0.72	1.29	3.89*	0.29	1.03	1.32	3.78*
Nod	0.49	0.81	1.73	5.38**	0.96	1.73	3.32	6.32**
Oral	0.63	0.28	0.26	3.64*	0.56	1.16	1.42	4.70**
Kick	0.23	0.20	0.10	0.55	0.30	0.22	0.30	0.21
Total	1.62	2.21	3.38	3.58*	2.11	4.14	6.36	5.55**

Conclusions

This study, therefore, confirms that access to high fibre at concentrate meal times reduces stereotypic behaviour. It also suggests that dividing the stabled horses concentrate ration into a number of smaller meals may be an effective means of reducing stereotypic oral behaviour in stabled horses, but that pre-feeding stereotypies may persist and that the practice may increase the frequency of stereotypic behaviour on unfed horses in visual contact.

References

Cooper, J.J, McDonald, L and Mills, D.S. (2000). The effect of increasing visual horizons on stereotypic weaving: implications for the social housing of stabled horses. *Applied Animal Behaviour Science*, **69**: 67-83.

McGreevy, P.D, and Nicol, C.J. (1998). The effect of short term prevention on the subsequent rate of crib-biting in Thoroughbred horses. *Equine Veterinary Journal Supplement*, **27**: 30-34.

Nicol, C.J., Davidson, H.P.D., Harris, P.A., Waters, A.J. and Wilson, A.D. (2002). Study of crib-biting and gastric inflammation and ulceration in young horses. *The Veterinary Record*, **151**: 658-660.

Analyses of the moisture content of hoof horn from horses, donkeys and laminitic donkeys

L Hopegood[1], S N Collins[2], R J Latham2, J D Reilly[3]
[1] The Nottingham Trent University, School of Land-based Studies, Brackenhurst, Southwell, Notts, NG25 OQF, UK. Email: lyn.hopegood@ntu.ac.uk
[2] De Montfort University, Faculty of Health and Life Sciences, The Gateway, Leicester, LE1 9BH, Email: rjlatham@dmu.ac.uk, Email: scollins@dmu.ac.uk
[3] Royal Army Veterinary Corps, Defence Animal Centre, Melton Mowbray, LE13 OSP

Introduction

The moisture content of keratinous materials such as hoof horn is important as the presence of moisture has an inverse relationship on the mechanical properties of hoof horn and may have a subsequent effect on the function of the hoof. Methods previously used to dehydrate samples to calculate the moisture content of hoof horn vary considerably (Hopegood, 2002). Subsequent comparison of results is therefore unreliable. A comparison of different methods of dehydrating hoof horn was therefore carried out to establish a standardised protocol for dehydrating hoof horn to assess its moisture content. The moisture content of donkey hoof horn from normal animals and those with laminitis has not been reported. Maclean (1971) established that the moisture content of cattle suffering from laminitis was significantly higher than normal hooves. The resultant standardised protocol from the first part of this study was then used to compare the moisture content of hoof horn samples taken from horses, donkeys and those donkeys that had suffered from laminitis.

Materials and methods

Samples for comparison of different methods of dehydrating donkey hoof horn to ascertain the moisture content were taken from clippings removed from the left forelimb of 31 donkeys during regular farrier treatment at The Donkey Sanctuary, Sidmouth, Devon. Samples were immediately wrapped to prevent moisture loss. Nine adjacent samples, encompassing the full hoof wall depth (HWD), were removed from the Midline Dead Centre (MDC) of the clipping. Samples were randomly allocated to one of the drying methods: room temperature, vacuum drying, freeze-drying, drying over phosphorus pentoxide, oven drying at 90, 100, 105, 110, 120°C. Samples were weighed daily until they established equilibrium mass. The moisture contents were calculated with reference to the original mass of samples. Samples of horse hoof horn were obtained from the clippings removed from the left forelimb of 16 unshod animals at the Friends of Bristol Horses Society. The laminitic samples of donkey hoof horn were obtained from the left fore hoof capsules from 15 donkeys from The Donkey Sanctuary that had been euthanased for reasons other than this study. Samples were removed from the MDC of the capsules at 50% hoof wall height. One full HWD sample was removed from the horse and laminitic donkey hoof samples. All samples were then dehydrated over phosphorus pentoxide.

Results

The moisture contents of samples dried by the different methods varied between 250-360 g/kg. The results for each technique were significantly different to those from at least 4 other methods ($p<0.05$, Mann-Whitney U tests). As samples dried over phosphorus pentoxide avoided both the unknown effect of high and low temperatures on hoof horn, showed a normal distribution and low variability (Coefficient of Variation (CV) of 6%) compared to other results (ranging 4-44%) this technique was then used to compare the moisture contents of further samples of hoof horn.

There were significant differences between mean moisture contents from each data set (ANOVA, $p<0.05$). The moisture content for horse hoof horn was significantly lower than that for both donkey clippings and from laminitic hoof horn (Tukey Test, $p<0.05$) (Figure 1). There was no significant difference between the moisture content of donkey clippings and laminitic hoof horn (Tukey Test, $p<0.05$).

Figure 1.
Comparison of
moisture contents.

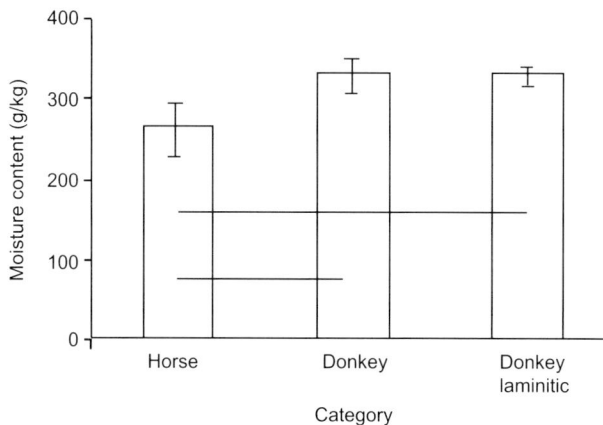

Error bars - standard deviations
———— significant difference ($p<0.05$)

Conclusions

A comparison of results of moisture contents of hoof horn in the literature should not be carried out unless the same drying techniques are used as different drying techniques result in different moisture contents. The moisture contents of donkey hoof horn, for both clippings (332 g/kg) and laminitic samples (332 g/kg) are higher than for horse hoof horn (264 g/kg). This may have implications for the practical management of both donkey and horse hoof horn and may be the reason why donkeys appear to suffer particular hoof problems. Laminitis may not affect the moisture content of hoof horn samples although it should be borne in mind that different sample sites were used.

References

Hopegood, L. (2002). *Tubule Density, Moisture Content and Mechanical Properties of Donkey Hoof Horn*. PhD Thesis, De Montfort University.

Maclean, C.W. (1971). The long term effects of laminitis in dairy cows. *Veterinary Record*. **89**: 34-37.

Tubule density at different sample sites around the *Stratum medium* of horse hoof horn

L Hopegood, D Abraham and C Shervey
The Nottingham Trent University, School of Land-based Studies, Brackenhurst, Southwell, Nottinghamshire, NG25 OQF, UK, Email: lyn.hopegood@ntu.ac.uk

Introduction

Microscopic analysis shows that equine hoof wall is composed of distinct phases of tubular and intertubular horn. The tubule density (TD) of hoof horn is found by ascertaining the number of tubules per unit area and is believed to relate to mechanical properties, resistance to wear, "quality", hardness and moisture content. The straight grid previously used to ascertain the TD of hoof horn may omit certain tubules on both the inner and outer hoof wall as the wall itself is curved. This present study examined whether there was a significant difference in results for TD between using the conventional straight grid method and a novel method using a curved grid. TD was also compared across three sample sites. An analysis of TD from four equal sections (zones) across the hoof wall was also carried out.

Materials and methods

The left fore feet of 10 randomly selected slaughterhouse horses and ponies were used to provide hoof samples. *Study 1* - Hoof sections were cut on a microtome from the midline dead centre (MDC) of the *Stratum medium*, were stained with Haematoxylin and Eosin and mounted on 35 mm slides and projected onto a screen. Curved grids were constructed for individual samples based on the curve of the outer *Stratum medium*. The chosen grid was placed over the projected image. A tubule count provided the Initial Tubule Density. A conversion was then applied to determine actual tubule density. *Study 2* - Samples were removed from the *Stratum medium* of the hoof wall at 3 sample sites: the MDC, medial and lateral sites. The method used in Study 1 to ascertain TD was also used for this study but the same straight grid was used for all samples.

Results

Study 1 There was no significant difference between the mean TD for the full hoof wall depth for samples analysed using the straight grid

(18 tubules mm^{-2}) and those analysed using the curved grid (19 tubules mm^{-2}) (Mann Whitney U test, p>0.05). However, the pattern of tubule distribution did not relate to the stepwise distribution described by previous authors for pony and horse hoof horn (Reilly et al., 1996; 1998) but did appear to be similar to that previously shown for donkey hoof horn by Hopegood (2002). However conflicting results may be due to samples being used from both horse and pony hoof horn.

Study 2 - The median tubule densities at the different sample sites were 6, 10 and 10 tubules mm^{-2} for medial, MDC and lateral samples respectively and are shown in Figure 1. There were significant differences in TD between the 3 sample sites (Kruskal-Wallis test, p<0.05). The TD for the medial site was significantly lower than either for midline or lateral samples (Mann Whitney U test, p<0.05). There was also a significant difference between TD for zonal comparisons between the three sample sites (Mann Whitney U test, p<0.05). The TD at different sample sites around the hoof wall capsule had previously not been ascertained.

Figure 1.
Tubule density at different sample sites around the hoof wall.

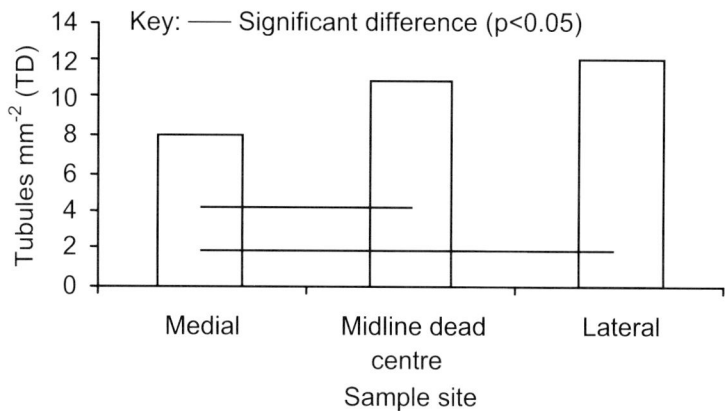

Conclusions

A straight or curved grid may be used to ascertain the TD of horse hoof horn. Studies on the TD of hoof horn should take into account the sampling site as TD varies according to sample position. As TD is believed to relate to the mechanical properties of the hoof, the establishment of a significantly lower TD for the medial sample site when compared to the other sites may reflect differences in the mechanical and functional properties of horse hoof horn in these different areas. Indeed, the medial wall of the hoof is often found to be more upright than the lateral wall. The TD may then be a reflection of the weight bearing capacity of the medial aspect of the wall.

References

Hopegood, L. (2002) *Tubule Density, Moisture Content and Mechanical Properties of Donkey Hoof Horn*. PhD Thesis, De Montfort University.

Reilly, J.D., Cottrell, D.F., Martin, R.J. and Cuddeford, D. (1996) Tubule density in equine hoof horn. *Biomimetics* 4:1, 23-35.

Reilly, J.D., Collins, S.N., Cope, B.C., Hopegood, L. and Latham, R.J. (1998) Tubule density of the *Stratum medium* of horse hoof. *Equine Veterinary Journal Supplement,* 26, 4-9.

Study of crib-biting and gastric ulceration and mucosal inflammation in foals

A J Badnell-Waters[1], A D Wilson[1], P A Harris[2], H B Davidson[2] and C J Nicol[1]

[1]Department of Clinical Veterinary Science, University of Bristol, Langford BS40 5DU, UK

[2]Equine Studies Group, Waltham Centre for Pet Nutrition, Waltham-on-the-Wolds, LE14 4RT, UK.

Address correspondence to Amanda Badnell-Waters, e-mail: amanda@badnellwaters.evesham.net

Introduction

Crib-biting is a stereotypic behaviour performed by approximately 5% of captive domestic horses. Dietary factors have been strongly associated with the development of oral stereotypies and risk factors for crib-biting, identified in recent epidemiological studies, include feeding high concentrate and/or low forage diets (Waters et al., 2002). Experimental work has shown that such diets are likely to result in increased gastric acidity (Murray and Eichorn, 1996; Nadeau et al., 2000). We therefore propose that young horses initiate crib-biting in an attempt to produce alkaline saliva to buffer their stomachs when alternative opportunities for mastication are limited. The aim of this study was to determine whether there was an association between crib-biting behaviour and stomach condition in foals.

Materials and methods

Foals that had recently started to perform crib-biting were recruited into the study and compared with non-stereotypic foals. The stomachs of 15 crib-biting foals and 9 normal foals were examined using a video endoscope. Foals were then randomly allocated to a control or an antacid diet for a 3-month period. Behaviour was monitored by direct observation throughout the 3-month period, and foals re-endoscoped at the end. Videos were scored blind by an independent observer.

Results

Crib-biting foals had significantly more inflamed, dry and ulcerated stomachs than normal foals on first examination (Mann-Whitney: $U = 36$; $N1 = 15$; $N2 = 9$; $p < 0.05$). Their stomachs also lacked normal folding on first examination, and were significantly smoother when re-examined (Mann-Whitney: $U = 11.5$; $N1 = 12$; $N2 = 7$; $p < 0.01$). Foals that received the antacid diet had fewer ulcers at the end of the trial (Mann-Whitney: $U = 27.5$; $N1 = 13$; $N2 = 9$). Most foals showed a reduction in crib-biting following treatment but the reduction was more pronounced in foals that received the antacid diet, and reduction in crib-biting correlated with reduction in ulceration (Rank correlation: $r = 58$; $N = 12$; $p = 0.05$).

Conclusions

We suggest that the stomachs of crib-biting foals were exposed to more acid conditions or were more sensitive to normal acidity levels than the stomachs of normal foals. The results of this study support the hypothesis that the initiation of oral stereotypy can be a response to a disturbance of the normal digestive process.

References

Murray, M.J. and Eichorn, E.S. (1996). Effects of intermittent food deprivation, intermittent food deprivation with Ranitidine, and stall confinement with free access to hay on gastric ulceration in horses. *American Journal of Veterinary Research.* 11: 1599-1603.

Nadeau, J.A., Andrews, F.M., Mathew, A.G., Argenzio, R.A., Blackford, J.T., Sohtell M., Saxton, A.M. (2000). Evaluation of diet as a cause of gastric ulcers in horses. *American Journal of Veterinary Research*, 61: 784-790.

Waters A.J; Nicol C.J. and French N.P. (2002). Factors influencing the development of stereotypic and redirected behaviour in young horses: the findings of a four year prospective epidemiological study. *Equine Veterinary Journal* 34 (6): 572-579.

The relationship between pre weaning social behaviour and racing performance in Thoroughbred horses

H Randle[1], T Counsell[2] and K Wilton[1]
[1]Seale-Hayne Faculty, University of Plymouth, Devon, TQ12 6NQ, UK, Email: h.randle@plymouth.ac.uk
[2]Duchy College, Stoke Climsland, Callington, Cornwall, PL17 8PB, UK, Email: tessa.counsell@cornwall.ac.uk

Introduction

Vast sums of money are invested in training racehorses, yet race earnings rarely cover training costs. The ability to accurately predict potential racing performance prior to training would be advantageous. Although a link between equine behaviour and later racing success has been suggested, no single measure has yet been identified, mainly due to the complex relationships between behavioural attributes, particularly those of a social nature. The influence of gender is also unclear. According to The Mental Health Model of Sport Performance a degeneration in mental state is reflected by a decline in performance. The success of over 70% of human athletes was identified using behavioural measures of personality. Reportedly equine 'winners' are 'calm' whilst 'losers' are 'aroused'. Such equine 'reactivity' ('temperament') has traditionally been investigated using contrived experiments. This study aimed to carefully examine equine social activity by exploring multiple aspects of social behaviour whilst avoiding the restrictive concept of dominance. The objective was to investigate the pre weaning social behaviour (PWSB) exhibited by Thoroughbred (TB) foals to identify potential measures for predicting future racing performance. Since it was not possible to link PWSB with racing performance in the time frame, in the interim the relationship between PWSB and their dams previous racing performance was explored.

Materials and methods

Ten TB foals (age 11-30d; 6 males, 4 females) were observed for 6.5 h/d for a period of 5d at The National Stud, Newmarket. The foals were observed from the outside of 3 adjacent 1.6ha paddocks each containing 3 or 4 mare-foal pairs, by an observer who had no knowledge of the mares previous performance. A Scan Sampling method was used, where the activity of each foal was recorded instantaneously every 15 minutes. Activity recorded included maintenance activity, type of social behaviour, if the foal was a Performer (P) or Receiver (R) and foals proximity to dam and nearest neighbour. A series of social activity measures were derived from the raw social and proximity data: %social, P Tendency, R Tendency, P/R ratio, Total Social Involvement (TSI), popularity (POP - as a neighbour) and Total Sociability Score (TSS=TSI+POP). Independent t tests were undertaken to investigate the effect of gender. A Principal Components Analysis (PCA) was conducted to examine the relations between the derived measures of social and proximity behaviour. Further correlations were undertaken to explore the relationship between foal behaviour and the dams previous racing performance (based on her average winnings).

Results

The foals were observed for a total of 312.5 equine hours. Time budgets illustrating maintenance behaviour demonstrated that only 0=8.39"1.25s.e.m. of the foals time was spent on social behaviour. Table 1 lists the individual scores for the derived social activity measures. Gender did not have a significant effect on any of these measures (P>0.05). The typical distance maintained between mare and foal was between 1-2m, although individual foals varied from 2-10m to <1m. The PCA revealed two Principal Components (PC1 and PC2). An Informal Scree Analysis of the Eigenvalues (where EV>1) revealed two Principal Components (Table 2). Inspection of the constituent variables indicate that PC1 reflects 'general sociability' whilst PC2 reflects 'tendency to be a performer rather than a receiver'. There were no significant relationships between foal behaviour and dams previous performance.

Table 1. Individual foal scores.

Foal	P	R	P/R	TSI	POP	TSS
1	27	15	1.8	42	31	73
2	8	3	2.7	11	5	16
3	19	8	2.4	27	33	60
4	50	11	4.6	61	20	81
5	19	8	2.4	27	19	46
6	23	8	2.9	31	20	51
7	20	13	1.5	33	21	54
8	12	3	4.0	15	2	17
9	33	2	16.5	35	0	35
10	15	3	5.0	18	2	20
Mean	22.6	7.4	4.28	30.0	15.3	45.3
s.e.m	3.79	1.45	1.40	4.58	3.86	7.28

| | Principal component | |
Variable	PC1 (EV=4.31)	PC2 (EV=1.93)
% social	-0.348	-0.276
P tendency	-0.375	-0.413
R tendency	-0.401	0.330
P/R	0.088	-0.671
TSI	-0.438	-0.237
POP	-0.385	0.386
TSS	-0.480	0.046

Table 2. Elements of principle components.

Conclusions

There were clear differences in the social behaviour attributes between individual foals even though they were very young. Sociability and tendency to be a Performer at the pre weaning stage were identified as being especially important attributes (as evidenced by social interactions *per se* and proximity relations). These attributes may provide useful measures for the prediction of future racing performance at two and three years old. In this group of foals pre weaning social behaviour was not related to the dams previous performance.

References

More, S. J. (1999). A longitudinal study of racing Thoroughbreds: performance during the first years of racing. *Australian Veterinary Journal* 77: 22.

Nicol, C. J. (2002). Equine learning: progress and suggestions for future research. *Applied Animal Behaviour Science* 78: 193-208

Raglin, S. J. (2001). Psychological factors in sport performance – the mental health model revisited. *Sports Medicine* 31: 875-890.

The effect of odour on the autonomic status of the horse

N Boyle, L Clarke and R Geering
Department of Agricultural Sciences, Imperial College London (Wye Campus), Wye, Ashford, Kent, TN25 5AH, UK, Email, r.geering@imperial.ac.uk

Introduction

It is well established that the equid olfactory system is highly sensitive. It has been suggested that there is an intimate link between the sense of smell in the horse (Williams, 1995) and the fight and flight response, which is mediated by the autonomic nervous system (Marlin and Nankervis, 2002), thus affecting the heart. In humans, it has been demonstrated that lavender can have a pronounced impact on heart rate variability (Saeki, 2000) but it is not yet know whether the autonomic nervous system of the horse can be influenced by odour. It is the aim of the study to determine the effect of odour on the autonomic status of the horses.

Materials and methods

Eight geldings were used in the study and the electrical activity of their heart was assessed using the "Polar S810" telemetric system before and during the presentation of either pig faeces or lavender to the horses' nostrils. The experiment was conducted over a two-day period so that their response to both odours could be determined. Data was analysed using "Research Tools" from Reynolds medical, which is a commercially available computer program, to generate time domain and frequency measures of heart rate variability. Statistical difference between the treatment groups was assessed using a balanced General Linear Model, ANOVA.

Results

Pig odour did not appear to have any affect on heart rate variability, primarily due to considerable variation in the data (Table 1). In contrast, lavender scent resulted in a reduction in the standard deviation of the mean of qualified NN intervals (SDNN) coefficient of variance ($P<0.05$), Low frequency power (LF) ($P<0.01$), LF/High frequency (HF) ratio ($P<0.01$) and Total Power ($P<0.01$).

Table 1.
The effect of
lavender or pig
odour on time
and frequency
domain spectral
analysis
parameters.

	Resting	Lavender	Resting	Lavender
SDNN Coeff. (msec)	83±10	65±11	82±10	70±9
LF (sec^2/Hz)	111±19	47±10	111±19	194±69
LF/HF ratio	8.8±2.1	5.2±1.4	8.8±2.2	12.1±6.7
Total Power (sec^2/Hz)	126±21	49±11	127±20	300±69

Values are meansSEM)

Conclusions

The detection of odours by the horse can alter their physiological autonomic status as reflected by a change in heart rate variability. The data suggests that lavender may have a calming influence on geldings whilst the pig odour appears to have the opposite effect.

References

Marlin, D. and Nankervis K. (2002). The cardiovascular system. In: *Equine Exercise Physiology*, Edited by Nankervis K., Blackwell Science, Oxford, UK, pp. 55-73.

Saeki, Y. (2000). The effect of foot-bath with or without the essential oil of lavender on the autonomic nervous system, a randomised trial. *Complementary Therapeutic Medicine* **89**(1): 2-7.

Williams, M. (1995). *Horse Psychology*, J. A. Allen, London, UK, pp. 61-62.

The effect of mirrors for the control of stereotypic weaving in the stabled horse

E Grime[1], A Northrop[1] M E Rosbotham[2], and K L Howells[2]
Equine Science[1] and Animal Studies[2] Department, Myerscough College, Bilsborrow, Preston PR3 0RY, UK
Email: KHowells@myerscough.ac.uk

Introduction

The exhibition and aetiology of abnormal behaviour in the equine is a coping response to a sub-optimal environment (McBride and Cuddeford, 2001). Physical prevention of performance of these behaviours could however lead to compromised welfare so alternative welfare enhancement methods have been sought as alternatives. Nicol (1999) indicated that the cause of weaving as a specific stereotypic behaviour is related to the management of the stabled horse. Enrichment devices have been successful in reducing abnormal behaviour and fostering increased expression of more normal behaviours in many animal species but studies have remained limited in relation to the equine. By strategic placement of mirrors in the stable the current study aimed to reduce the stabled horses need to perform stereotypic weaving behaviour without the use of physical restraint in order to enhance the welfare outcome.

Materials and methods

Four horses, two males and two females were housed individually in 10*12m loose boxes and provided with straw bedding. Each horse had been known to exhibit weaving behaviour for at least two years prior to trial commencement and horse-to-horse contact between loose boxes was prohibited. One 1*1m Lincoln Stable mirror was placed in each of the loose boxes adjacent to the stable door (0.5m in) on the right hand side 1.5m from ground level. Pre-enrichment, Enrichment and Post-enrichment observation phases were undertaken each with duration of two weeks. Daily behavioural observations were planned to coincide with routine feeding times as this has been shown to be when weaving behaviour frequencies are intensified (Cooper, McDonald and Mills, 2000). Three observation periods per day Monday to Friday were taken during each phase. The scan sampling method of observing behaviour was incorporated into a check sheet template with each horse's behaviour being scanned and recorded at 30-second intervals. Specific behaviours were defined by an ethogram adapted from McAfee, Mills and Cooper, (2002).

Results

A significant difference was found between the pre-enrichment duration and frequency of weaning behaviours and those during the enrichment phase (P<0.001). Head shaking and door kicking behaviours were also significantly reduced (P<0.001) during the enrichment phase, as presented in Table 1. Ingestion of hay increased and dozing behaviour increased significantly (P<0.001) during the enrichment phase in comparison with the pre-enrichment phase. Weaving behaviours, head shaking and ground pawing increased significantly in the post-enrichment phase when compared with the pre-enrichment phase (P<0.001). Ingestion of hay and dozing behaviour was reduced in occurrence in the post-enrichment phase comparative to the pre-enrichment phase.

Table 1.
Scans of behaviours (%) for all horses throughout the three enrichment phases.

Behaviour	Pre-enrichment (Mean ± s.d.)	Enrichment (Mean ± s.d.)	Post-enrichment (Mean ± s.d.)
Weaving	31.2 ± 11.4	14.5 ± 10.0	27.7 ± 9.0
Head shaking	3.0 ± 3.2	0.5 ± 0.9	3.1 ± 3.1
Pawing ground	3.0 ± 3.6	0.9 ± 1.4	2.0 ± 2.9
Ingest Hay	36.9 ± 10.1	41.8 ± 10.4	38.8 ± 9.0
Dozing	16.5 ± 12.0	29.8 ± 15.1	18.5 ± 10.1

Conclusions

Weaving and other equine specific stereotypic behaviours can be reduced in occurrence with the introduction of mirrors into the stable environment.

References

Cooper, J. J., McDonald, L. and Mills, D. S. (2000). The effect of increasing visual horizons on stereotypic weaving: Implications for the social housing of stabled horses. *Applied Animal Behaviour Science* **69**: 67-83.

McAfee, L. M., Mills, D. S. and Cooper, J. J. (2002). The use of mirrors for the control of stereotypic weaving behaviour in the stabled horse. *Applied Animal Behaviour Science* **78**: 159-173.

McBride, S. D. and Cuddeford, D. (2001). The putative welfare reducing effects of preventing equine stereotypic behaviour. *Animal Welfare* **10**: 173-189.

Nicol, C. (1999). Understanding equine stereotypies. *Equine Veterinary Journal* **28**: 20-25.

Comparative methods of preventing stereotypic behaviour in horses

E Lebatteux[1], M E Rosbotham[1], A Northrop[2] and K L Howells[1]
*Animal Studies[1] and Equine Science[2] Departments, Myerscough College, Bilsborrow, Preston PR3 0RY, UK
Email: KHowells@myerscough.ac.uk*

Introduction

Over fifteen per cent of domesticated horses exhibit a form of stereotypic behaviour (Luescher, McKeown and Halip, 1991). These behaviours can represent or develop into health and welfare disequilibrium and crib-biting and weaving are two of the most commonly performed behaviours of this type. As these behaviours have been described as having no obvious function (Mason, 1991), procedures concerned with their ultimate prevention have been promoted within the equine industry. The aim of this study was to review the success of the physical, pharmacological, dietary, environmental and surgical methods available to horse owners for the prevention of equine stereotypies.

Materials and methods

A total of 300 questionnaires were distributed between riding schools, trekking centres and livery yards throughout the Northern counties of England along with horse owning BSc Equine students at Myerscough College, Lancashire. Of the total, 200 were distributed by post and the remainder researcher administered. The survey attempted to establish the environmental (stable design, extent of visual contact with other horses), behavioural (stereotypies exhibited) and nutritional (forage/concentrate dietary inclusion) aspects of each horse owned by the respondents and resulted in profiles for 225 equine subjects. A concluding section of the questionnaire covered the methods of prevention of stereotypic behaviours that respondents had attempted or were employing currently with horses that were exhibiting stereotypic behaviour.

Results

The response rates for the postal-distributed survey and the researcher-administered questionnaire were 41% and 60% respectively. From these responses 23.1% of horses were reported to exhibit stereotypic behaviour presently: 10% crib-biting behaviour, 9.3% weaving behaviour and 3.1% wood-chewing behaviour. In addition 16.9% of horses were reported to have exhibited stereotypic behaviour in the past having recovered. From total respondents, none reported use of pharmacological therapy or surgical methods as means of stereotypy prevention while increasing grazing, forage and exercise were reported partially successful. There was a significant relationship ($P<0.05$) between weaving/crib-biting and visual contact with other horses as presented in Table 1. Survey responses were analysed with SNAP survey software and the effect of preventative method was assessed by *t*-test and Spearman Rank correlation.

Table 1. Effects of visual contact, tactile contact, provision of straw as bedding and the provision of hay as main forage on the incidence of weaving and crib-biting behaviours (Spearman Matrix).

	Weavers		Crib-biters	
	Correlation	Probability	Correlation	Probability
Visual contact	-0.3517	0.0096	-0.3308	0.0141
Tactile contact	-0.1054	0.2479	0.1839	0.1161
Provision of straw bedding	-0.0559	0.3593	0.0279	0.4288
Provision of hay as main forage	-0.0091	0.4766	-0.0204	0.4477

Conclusions

Physical, pharmacological and surgical methods were not reported as popular choices or adequately successful means of preventing stereotypic behaviour. The results of the present study demonstrate however, that as the level of visual contact is increased between horses the incidence of both crib-biting and weaving is significantly reduced ($P<0.05$). The results are in agreement with Cooper, McDonald and Mills (2000) in showing that a combination of increasing forage levels, tactile contact and more notably visual contact can allow the horse's natural motivations to be fulfilled.

References

Cooper, J. J., McDonald, L. and Mills, D. S. (2000). The effect of increasing visual horizons on stereotypic weaving: implications for the social housing of stabled horses. *Applied Animal Behaviour Science* **69**: 67-83.

Lueschers, U. A., McKeown D. B. and Halip, J. (1991). Reviewing the causes of obsessive-compulsive disorder in horses. *Veterinary Medicine* **86**: 527-531.

Mason, G. (1991). Stereotypies: a critical review, *Animal Behaviour* **41**: 1015-1037.

Oxidative DNA damage in equine peripheral blood mononuclear cells (PBMC) induced by acute natural allergen challenge

D A Kingston[1], D J Marlin[1], N C Smith[1], C M Deaton[1], K Saunders[2], J Kydd[2], and P A Harris[3]

Centres for Equine Studies[1] and Preventive Medicine[2], Animal Health Trust, Newmarket, UK, and Equine Studies Group [3], Waltham Centre for Pet Nutrition, Waltham-on-the-Wolds, UK

Introduction

Oxidative stress occurs when antioxidant defence mechanisms are overwhelmed by free radicals and may lead to damage to DNA, which has been implicated in processes such as ageing and cancer. The Comet assay allows detection of oxidative DNA damage in individual cells. As horses with recurrent airway obstruction (RAO) have been shown to demonstrate low antioxidant status and oxidative stress, we hypothesised that peripheral blood mononuclear cells (PBMC) of horses with RAO would demonstrate increases in DNA damage following natural allergen challenge.

Materials and methods

Six horses (mean age 15 years, range 8-23 years) diagnosed with RAO (in remission) and 6 healthy breed matched controls (mean age 9 years, range 5-15 years) were studied. Blood samples were collected 7 days prior to challenge and immediately and 3 days after stabling on mouldy hay and straw for 24h. All animals were kept at grass prior to and after the challenge period. Bronchoalveolar lavage (BAL) was performed and neutrophil counts determined. Venous blood samples (15ml) were placed into plain glass tubes containing 0.5ml 15u/ml preservative free heparin in phosphate buffered saline (PBS). PBMC were isolated by gradient centrifugation. Comet assay and analysis by visual scoring were carried out according to Heaton, P.R., Ransley, R., Charlton, C., Mann, S.J., Stevenson, J., Smith, B.H., Rawlings, J.M. and Harper, (2002). Endogenous damage (endo) and susceptibility to oxidative damage following hydrogen peroxide incubation (exo) were determined in each sample at each time point.

Results

Immediately after stabling, RAO-affected horses had an increased number of neutrophils in BAL fluid (from 10 ± 4 /μl to 225 ± 136 /μl;

189

mean±sd), which declined to 35±29 /μl by 3 days post-exposure. Control horses demonstrated a smaller increase in neutrophil number following stabling (28±21 /μl at 24h from 10±4 /μl at −7day) compared to RAO-affected horses (P=0.004). There were no differences in mean endo or exo oxidative DNA damage between the control (endo=38±7 arbitrary units, AU; exo=59±11 AU) and RAO-affected groups (endo=37±9 AU; exo=71±21 AU), prior to stabling and neither group demonstrated an increase in endogenous or exogenous DNA damage after challenge.

Conclusion

In conclusion, acute natural allergen challenge does not induce oxidative DNA damage or increase the susceptibility to damage of circulating PBMC in either RAO-affected horses or non-RAO controls.

References

Heaton, P.R., Ransley, R., Charlton, C., Mann, S.J., Stevenson, J., Smith, B.H., Rawlings, J.M. and Harper, E.J. (2002) Application of single-cell gel electrophoresis (Comet) assay for assessing levels of DNA damage in canine and feline leukocytes. *Journal of Nutrition*, **132**: 1598S-1603S.

A comparative analysis of population characteristics of the competing progeny (within UK) of UK elite eventing sires and all competing event horses within the UK

T C Whitaker and J Hill
Faculty of Applied Science, Writtle College, Chelmsford, Essex, UK, Email: tcw@writtle.ac.uk

Introduction

A greater understanding of the population characteristics of sport horse populations is required to enable potential breed improvement programmes to be formulated correctly and be effective in their outcomes. To date limited research has been conducted into the UK sport horse population.

Materials and methods

A selected group of progeny (n=339) sired by elite eventing stallions was examined. In the context of this study elite sires were defined as those that were ranked 1-10 by total lifetime points won by competing progeny up to the end of 2000 (British Horse Database, 2000). Comparative analysis was undertaken between the selected group and all competing eventing horses in 2000 (n=9387) (British Horse Database, 2000). Data collected for both groups included, total lifetime points won at eventing and dressage and total lifetime money won at show jumping. Basic descriptive statistics were produced for each data set (Table 1). Product moment correlations were performed for all discipline areas (Table 2). Data transformation was applied using LOG+1(Hassenstein, Roehe, and Kalm, 1996).

Results

The mean progeny score for the elite data set for eventing points attained was +18.77 points when compared to the whole population. This trend was repeated for dressage at +28.19 point for the elite group. The elite group performance was however poorer at show jumping –£23.99, a large standard error was however apparent. Very large coefficience of variations were observed in all data sets. It is however apparent that the whole population data set displays larger coefficience of variation for the three performance areas. Product moment correlations for the whole population showed effectively no correlation between eventing and dressage, and dressage and show jumping scores. A slight positive correlation was observed between eventing and show jumping. The elite population effectively showed no correlation between eventing and dressage. In contrast to the whole population a negative correlation was observed between show jumping and dressage for the elite group. A weaker positive correlation was observed between eventing and show jumping within the elite group.

Table 1. Descriptive population statistics for elite group and whole population.

	μ	s.e.m	σ	V	Skew	Skew Log +1
Elite group						
Eventing	42.81	6.49	97.64	228	4.272	0.239
Dressage	68.15	7.62	85.88	126	2.089	-0.506
Show jumping	94.75	26.14	379.67	400	8.370	0.218
Whole population						
Eventing	24.04	0.78	75.41	313	8.35	0.787
Dressage	39.96	1.32	65.90	164	2.87	0.062
Show jumping	118.74	7.37	550.60	464	25.24	-0.139

Table 2. Person Product Moment Correlations for elite group and whole population.

	Eventing	*Show jumping*
Elite group		
Dressage	0.023 (n=76)	-0.312** (n=61)
Eventing		0.148* (n=139)
Whole population		
Dressage	0.022 (n=2493)	0.034 (n=1532)
Eventing		0.288** (n=5586)

* (P<0.05), ** (P<0.01), 1 tailed test

Conclusions

The mean lifetime progeny winnings of elite eventing sires is almost double that of the whole populations scores, additionally the variation within the elite population is lower for all disciplines. These factors support the opinion that elite sires are exerting an improving influence that is more consistent. A negative correlation is demonstrated amongst the elite eventing progeny that participate in dressage and show jumping. All other correlations are however weak. It has been demonstrated that pre-selection may play an important role in these types of correlations and careful consideration needs to be taken before drawing any conclusions (Philipsson, Arnason, and Bergsten, 1990).

References

British Horse Database (2000). *Sports Horse Annual 2001*. British Horse Database Ltd. Wellingbourgh, UK.

Philipsson, J. Arnason, T and Bergsten, K. (1990) Alternative selection strategies for performance of the Swedish warmblood horse. *Livestock Production Science* **24**: 273-285.

Hassenstein, C., Roehe, R and Kalm, E. (1996) Estimation of genetic parameters of German Riding Horses ranked by their competition place using Gibbs Sampling. *47th Annual Meeting of the European Association of Animal Production*, Lillehammer, Norway, 26-29th August 1996.

A comparative analysis of population characteristics of the competing progeny (within UK) of UK elite show jumping sires and all competing show jumping horses within the UK

T C Whitaker and J Hill
Faculty of Applied Science, Writtle College, Chelmsford, Essex, UK, Email: tcw@writtle.ac.uk

Introduction

The UK currently has limited success at producing show jumping horses of international standard (WBFSH, 2002). For improvement to occur a greater understanding of the population dynamics of the show jumping population is required

Materials and methods

A selected group of progeny (n=304) sired by elite show jumping stallions was examined. Elite sires were defined as those that were ranked 1-10 in the 2001 Sport Horse Annual (British Horse Database, 2000) by total lifetime earnings of progeny competing in 2000. Comparative analysis was conducted with all competing show jumping horses (n=22421) in 2000 (British Horse Database, 2000). Data collected comprised of total lifetime money won at show jumping, total lifetime points won at eventing and dressage. Descriptive statistics were produced for each data set (Table 1). Product Moment Correlations were performed for all discipline areas (Table 2). Data transformation was applied using LOG+1(Hassenstein, Roehe and Kalm, 1996) to account for the skewness in all data sets.

Results

The mean earnings of the elite sire group progeny was £750.44 more than the mean for the whole population. The elite show jumping group performed better at eventing +19.84 and slightly better at dressage +4.75 points than the whole show jumping population. Large standard errors and standard deviations were observed for show jumping performance within both populations. Very large co-efficience of variations were observed in all data sets, although these were greater in the whole population, standard errors of mean were however smaller in the whole population. Product Moment Correlations for the whole population showed very weak positive correlation between dressage and show jumping and eventing and dressage, a stronger correlation was observed between show jumping and eventing. The elite group displayed a negative correlation between show jumping and dressage. No effective correlations were observed between eventing and dressage performances and show jumping within the elite group.

<table>
	μ	s.e.m	σ	V	Skew	Skew Log + 1
Elite group						
Show jumping	1554.04	450.15	7230.47	465.27	9.06	-0.27
Dressage	59.24	9.72	77.16	130.25	1.34	-0.26
Eventing	29.26	11.34	69.92	117.52	3.38	0.84
Whole population						
Show jumping	803.58	61.17	9158.80	1139.66	148.47	0.03
Dressage	36.40	0.98	73.54	202.32	49.40	0.14
Eventing	24.51	1.26	62.43	254.71	74.57	0.66
</table>

Table 1. Descriptive population statistics for elite group and whole population.

<table>
	Show jumping	Dressage
Elite group		
Show jumping		-0.278 (n=22)
Eventing	0.035 (n=26)	0.065 (n=15)
Whole population		
Show jumping		0.037* (n=2448)
Eventing	0.288* (n=5586)	0.066** (n=1532)
</table>

Table 2. Person Product Moment Correlations for elite group and whole population.

* ($P<0.05$), ** ($P<0.01$), 1 tailed test

Conclusions

The mean lifetime earnings of the elite show jumping group is approximately double that of the whole population. The large standard errors of means, standard deviations and co-efficience of variations indicate that the use of lifetime earnings as a measure within the population is potentially unreliable and any conclusion drawn have to have this in mind. The elite group correlations, displayed small group sizes and no correlations produced significance, however show jumping and dressage displayed a negative correlation ($r=-0.278$). This may support the opinion of Schade, Bruns and Goldek (1994) that there is an antagonism between show jumping and dressage performance.

References

British Horse Database (2000). *Sports Horse Annual 2001*. British Horse Database Ltd. Wellingbourgh, UK.

Hassenstein, C., Roehe, R and Kalm, E. (1996) Estimation of genetic parameters of German Riding Horses ranked by their competition place using Gibbs Sampling. *47th Annual Meeting of the European Association of Animal Production*, Lillehammer, Norway, 26-29th August 1996.

Schade, Bruns, Goldek (1994) Genetic aspects of breeding for dressage or jumping ability in riding horse populations. *EAAP Commission Report 1994*.

WBFSH (2002), *2001/2002 WBFSH International Breeding Guide*. Best Communication and Management, The Netherlands.

The occurrence and outcome of multiple ovulation and pregnancy in the Thoroughbred mare

M C G Davies Morel and S Dennis
Institute of Rural Studies, University of Wales, Aberystwyth, Ceredigion SY24 5DP UK, Email: mid@aber.ac.uk

Introduction

Multiple pregnancies (MP) are almost always the result of multiple ovulations (MO) (Ginther and Bergfelt, 1988). Twinning is the most common non-infectious cause of abortion in the mare (Roberts and Myhre, 1983) and represents a significant economic loss. As a consequence, further investigation into its occurrence and outcome is warranted in order to determine effective management practices.

Materials and methods

Ultrasonic scanning was used to monitor and classify ovulation in 1582 Thoroughbred mares. Further ultrasonic scanning at Day 13 was used to identify and classify the resulting pregnancies (single pregnancy (SP), or MP)) in varying samples of mares. All MP mares underwent manual reduction of the smallest embryo at Day 13. 1170 mares, both SP and crushed MP, were monitored for the remainder of pregnancy with regard to pregnancy failure or success. Chi squared was used to test for significance throughout.

Results

The MO rate was 19.6% (19.0% double ovulation (DO), 0.6% triple ovulation (TO)). Bilateral DO (58.3%) were significantly ($p<0.01$) more frequent than unilateral DO (41.7%). 50.4%, 49.0% and 0.6% of the 339 pregnancies resulting from MO mares resulted in SP, DP and triple pregnancies (TP) respectively. In a sample of 230 pregnancies (170 SP, 60 unilateral DP) a significant ($p<0.05$ and $p<0.01$ respectively) preference for location of the conceptuses in the right horn was evident at Day 13 post ovulation (Table 1). Of the 1170 pregnancies monitored post Day 13, 14.4% and 18.2% of SP and crushed MP respectively failed to produce a live foal. No significant differences were evident between the ultimate failure of SP and crushed MP. However, significantly ($p<0.01$) more crushed MP failed due to early abortion (Day 13 – Day 150) than SP (6.6% vs 2.8%) and significantly (p<0.05) more SP failed due to late abortion (after Day 150) than crushed MP (2.9% vs 1.1%) (Table 2).

Table 1.
Frequency of location of D13 pregnancies (n = 230 pregnancies).

	Left Horn (%)	Right Horn (%)	Total
Single pregnancy	72 (42.4)[a]	98 (57.6)[b]	170
Unilateral DP	19 (31.7)[c]	41 (68.3)[d]	60
Total	91 (39.6)[c]	139 (60.4)[d]	230

[ab] significant p<0.05 difference.
[cd] significant p<0.01 difference.

Table 2.
The incidence of the possible outcomes of single pregnancy and manually reduced MP (n = 1170).

	Live foal n (%)	Barren* n (%)	Early abortion n (%)	Slipped foal¤ n (%	Dead foal n (%)	Foal died n (%)	Total pregnan-cies n (%)
Single pregnancy	847 (85.6)	53 (5.4)	28 (2.8)[a]	29 (2.9)[c]	16 (1.6)	16 (1.6)	989 (100)
MP (crushed)	148 (81.8)	9 (5.0)	12 (6.6)[b]	2 (1.1)[d]	6 (3.3)	3 (1.7)	181 (100)
Total	995 (85.0)	62 (5.3)	40 (3.4)	31 (2.6)	22 (1.8)	19 (1.6)	1170 (100)

*The term barren refers to the early reabsorption of pregnancy (prior to Day 13), figures for MP crushed in this instance refer to MO mares with evidence of failed MP at Day 13.
¤The term slipped foal refers to the birth of a dead foal before term (after Day 150)
[ab]Significant p<0.01 difference.
[cd]Significant difference p<0.05.

Conclusions

In conclusion a) embryo loss prior to and post Day 150 represents a significant loss in the manually reduced MP and SP respectively, b) significantly more conceptuses from unilateral DP and SP are located in the right uterine horn. The reason for these observations is unclear but they indicate times of significant risk to, and the probable location of, conceptuses and so aid in the monitoring and management of the Thoroughbred broodmare.

References

Ginther, O.J. and Bergelt, D. R. (1988). Embryonic reduction before day 11 in mares with twin conceptuses. *Journal of Animal Science* 66: 1727-1731.

Roberts, S.J. and Myhre, G. (1983). A review of twinning in horses and the possible therapeutic value of supplemental progesterone to prevent abortion of equine fetuses in the latter half of the gestation period. *Cornell Veterinarian* 73: 257-264.

Gender differences in cognition and perception may affect equine performance

J Murphy[1], T Waldmann[2] and S Arkins[1]
Departments of [1]Life Sciences and [2]Manufacturing and Operations Engineering, University of Limerick, Ireland, E-mail: jack.murphy@ul.ie

A competition horse: male or female?

Horses and their owners participate in an increasingly diverse range of equestrian pursuits including such activities as racing, show-jumping, endurance riding, carriage driving, dressage, hunting, pony club games, polo and leisure trekking. The majority of owners and riders within the disciplines of equitation appear to have a general preference toward using male horses as the chosen competition animal. Although not exclusively so, stallions and geldings are quite often physically bigger and stronger than fillies and mares and may enjoy some athletic advantage as a result. However, it is known from studies involving humans and small laboratory animals (mice and rats) that some gender differences in cognitive function may also affect performance where tasks involve 3-dimensional objects and elements of depth perception (Morris, Garrud, Rawlins, and O' Keefe, 1982). The horse has laterally placed eyes (Budiansky, 1997 and Figure 1) and therefore a stereoscopic (binocular) visual field of approximately 65-70°.

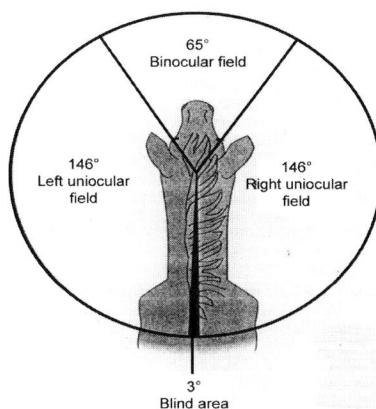

Figure 1.
Monocular and binocular visual fields in the horse.

Designing the test apparatus

We are currently investigating possible differences in cognitive function between male and female horses. In particular, we are attempting to get a better understanding of how horses actually perceive images of objects (like fences in show-jumping) and if male and female horses might deal with these perceptions differently. In order to conduct some of these investigations, we designed a unique test apparatus (Figure 2) to test the visuo-spatial ability of equine subjects by conducting a series of novel experimental tests. Visuo-spatial ability is a cognitive function (Farrell and Roberston, 2000) that allows individuals (humans, horses and other animals) to interpret distance and depth perception and keep track of the spatial environment – in other words it is how we know where we are!

Figure 2. Assessing a horse for visuo-spatial ability.

Setting up tests

The test apparatus is similar to a series of four stalls, each with a feed bin but also a barrier that either allows or prevents the animals access to the ration of feed concentrates in the feed bin. The horses were presented with a series of tests where they had to choose one of the four stalls in order to get access to the feed bin. During all tests, three out of four stalls had barriers positioned such that the horses were denied access to the feed bins. The horses were lead into starting position 9 metres from the test apparatus (Figure 3), were held in position there by a handler and allowed to observe the setting up of the test including the positioning of the barriers and the delivery of feed to the feed bins. The observers monitored the performance of the horses over the series of tests and recorded the length of time taken by each horse to access the feed bin. Some of the horses entered incorrect stalls (often more than once) and the number of such errors were also noted. A total of 62 animals, stallions (n = 1), geldings (n = 33) and mares (n = 28) have been put through their paces. Results to date have indicated that the male horses have performed significantly better (P<0.05) than the females during the series of tests.

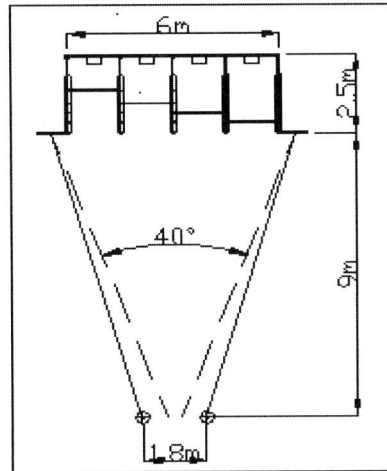

Figure 3.
Plan view of test apparatus and starting position.

Future equine research proposals

We are also involved in further research including an investigation into the issue of 'sidedness' and the incidence of preferential and innate idiosyncratic biomechanical gait movements in equines. Other topics also under investigation involve the limitations of monocular vision, spatio-temporal behavioural patterns in young foals and their possible impact on the future athletic performance of the horses.

References

Budiansky, S. (1997). *The Nature of Horses: Their Evolution, Intelligence and Behaviour.* Weidenfeld and Nicolson, London, UK.

Farrell, M.J. and Robertson I.H. (2000). The automatic updating of egocentric spatial relationships and its impairment due to right posterior cortical lesions. *Neuropsychologia* 38: 585-595.

Morris, R. G. M., Garrud, P., Rawlins, J. N. P. and O' Keefe, J. (1982). Place navigation impaired in rats with hippocampal lesions. *Nature,* 182: 681-683.

Gender differences in equine visuo-spatial ability

J Murphy[1], T Waldmann[2] and S Arkins[1]
Departments of [1]Life Sciences and [2]Manufacturing and Operations Engineering, University of Limerick, E-mail: jack.murphy@ul.ie

Introduction

Visuo-spatial ability has been defined as the mechanism that allows individuals to keep track of their changing spatial relationships as they move through their environment (Farrell and Robertson, 2000). In many species, moderate to large differences in visuo-spatial ability have been reported between genders, with superior visuo-spatial ability being reported in males of all species investigated to date (Masters and Sanders, 1993). There is an acknowledged preference among trainers and owners toward male horses in several disciplines of equitation. While this preference for male horses may be partially explained by physical differences, visuo-spatial ability is important in many aspects of equitation. The objective of this study was to determine if visuo-spatial ability differed between male and female horses.

Materials and methods

Visuo-spatial ability was investigated in a sample of 62 horses (males* = 34 and females = 28) at two equestrian centres using a novel test and test apparatus. The apparatus and test procedures required the horses to use elements of spatial awareness and the ability to differentiate elements of distance and depth perception to complete a series of six tests. The test apparatus was designed such that breastplate barriers controlled access by the horses to feed bins in all but one stall during each test (Figure 1). The barriers, which were in positions, 1.1, 1.5, 1.8 or 2.1 m from the feed bins, were changed randomly within stalls for each test. Selection of the correct stall option allowed the horse under test access to the feed bin and successful completion of individual tests, which resulted in food reinforcement, but errors and incomplete tests were not rewarded. On the evening prior to testing, the animals underwent a period of

* Males were geldings (n = 33) and stallions (n = 1).

familiarization with the test apparatus. Testing was conducted in the mornings just prior to normal feeding times. All relevant data were recorded and subjected to analysis of variance, chi-square for tests of association and t-tests for repeated measures for statistical analysis. The parameters under investigation during the tests were the subjects':

1) time required for completion of each test,
2) errors (incorrect choices) during tests,
3) number of successful test outcomes.

Figure 1.
Test apparatus.

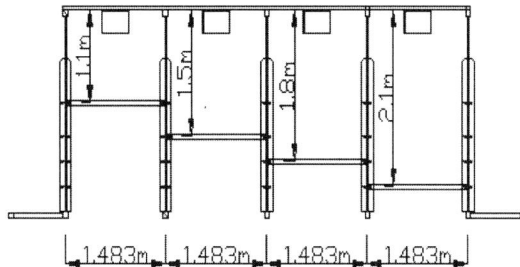

Results

While there were no differences in mean times for test completion between genders (ANOVA), 83.3% of horses that successfully completed all 6 tests in a time of 30 s or less per test were male (P<0.05). There were no differences in mean times for female subjects between any of the individual six tests (Figure 2). However, there were significant differences in mean times for male subjects between test 1 and test 2 (P<0.05), test 1 and test 3 (P<0.001), test 1 and test 4 (P<0.05) and test 1 and test 5 (P<0.05), suggesting that males improved over the course of the experiments.

Figure 2.
Comparison of subjects' mean completion times during all 6 tests in male and female horses.

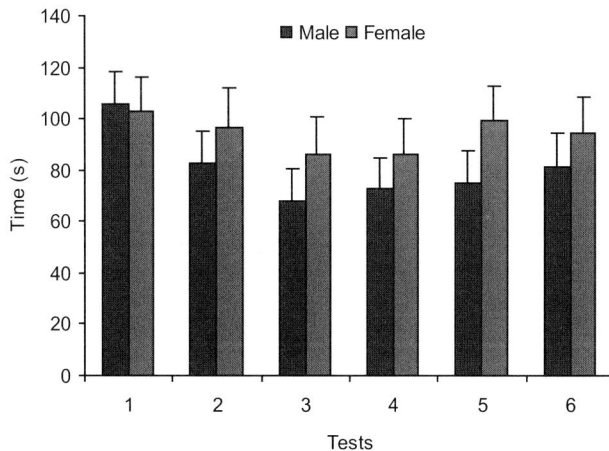

The differences between genders for the number of errors made during test execution approached significance (P = .054; Figure 3). Males had a lower mean total number of errors for all tests. There were also significant differences between genders in the ability to successfully complete all 6 tests (72% males vs. 28% females; P<0.05).

Figure 3.
Comparison of subjects' mean number of errors per test in male and female horses.

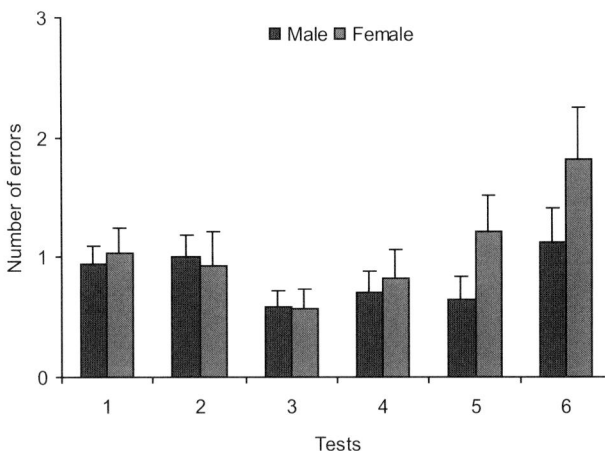

Conclusions

These results provide the first behavioural demonstration of the existence of superior visuo-spatial ability in male horses in line with similar results reported in other species. The results also appear to support some of the more recent equine research where (Flannery, 1997) has suggested that horses have the ability to rationalise and form concepts necessary to solve more complex and higher order discrimination problems.

References

Farrell, M.J. and Robertson I.H. (2000). The automatic updating of egocentric spatial relationships and its impairment due to right posterior cortical lesions. *Neuropsychologia* 38: 585-595.

Flannery, B (1997). Relational discrimination learning in horses. *Applied Animal Behaviour Science* 54: 267-280.

Masters, M.S. and Sanders, B. (1993). Is the gender difference in mental rotation disappearing? *Behavior Genetics* 23: 337-342.

An investigation into whether a significant correlation exists between feed pH and faecal pH of horses fed a variety of different diets

R Wheeler

Welsh Institute of Rural Studies. Llanbadarn Fawr, University of Wales Aberystwyth, Aberystwyth, SY23 3AL, UK

Introduction

Previous work has shown that an increase in readily fermentable carbohydrates (CHO) in the diet of pigs is reflected by a decrease in faecal pH (Canh *et al.*, 1998). It is well documented that a similar increase in CHO in the diet of equids is reflected by a decrease in caecal pH (McLean *et al.*, 2000). Little work has been undertaken to determine if such a relationship exists in equids or indeed if there is a correlation between dietary pH and faecal pH.

Materials and methods

The feed pH and faecal pH of six groups, each of 8 horses, receiving a variety of diets were examined. The dietary groups covered a wide range of feeds from purely forage based to predominantly concentrate based. Diet 1 consisted of 100% Hay (Hy), diet 2 of 100% grass (Gr), diet 3 of a 70:30 mix of Haylage (Hg) and grass (Gr): stud nuts, diet 4 of a 50:50 mix of haylage and competition mix, diet 5 a 30:70 mix of haylage to race mix and diet 6 a 60:40 mix of haylage, alfalfa and sugar beet pulp: rolled oats. Upon collection 10g of each feed sample were weighed in relation to the ratios in which they were fed, samples were than suspended in 100ml of distilled water. Similarly 10g of each faecal sample were weighed and suspended in 100ml of distilled water. A pH meter (Jenway 3051, Fisher Scientific, U.K.) was calibrated against a pH7 buffer at an approximate room temperature of 20°C. Each sample was then agitated for 30 seconds and a pH reading was taken. Results were recorded into Microsoft Excel (Millennium, 2000) for statistical analysis. Feed pH and Faecal pH were analysed using analysis of variance (ANOVA) with feed and faeces as factors using Genstat 6.1 (Lawes Agricultural Trust, 2002) Further analysis was undertaken to determine if a relationship existed between feed pH and faecal pH using regression analysis (Genstat 6.1, Lawes Agricultural Trust, 2002).

Results

The results of the study indicate that no apparent correlation between feed pH and faecal pH existed. However the study did reveal significant (P<0.05) differences in mean feed pH values between all dietary groups. Moreover both significant (P<0.05) and highly significant (P<0.01) differences were seen between faecal pH values.

Table 1.
Mean feed and faecal pH values calculated for the six dietary groups.

Diet	1	2	3	4	5	6	s.e.d.	sig
Feed pH	5.9[e]	6.36[f]	5.6[c]	5.3[a]	5.75[d]	5.53[b]	0.030	*
Faecal pH	6.66[b]	6.62[b]	6.86[b]	6.61[b]	6.8[b]	6.25[a]	0.172	*

[abc] Values within the same row not showing common superscripts differ significantly (P<0.05).

Conclusion

There was no apparent relationship between the amount of concentrate fed and a reduction in feed pH. However the effect of processing on the grain and the addition of buffering substances may have affected feed pH. The mean faecal pH for diet 6 which contained haylage, alfalfa, sugar beet pulp and rolled oats was significantly (P<0.05) lower than all other diets. Alfalfa and sugar beet pulp are both forage substances; these were meal fed along with the rolled oats in this diet. The addition of forage to the rolled oats may have decreased the MRT of the diet. However, diets 3, 4 and 5 all fed the concentrate portion of their diet as a straight feed. The size of meal fed may also have an effect on faecal pH. Starch intake should not exceed between 2 and 4g/Kg BW per meal in order to ensure that the capacity of the small intestine to absorb starch is not exceeded (Kienzle, 1994). A greater number of meals fed per day has been found to reduce the affect of starch influx on caecal pH and may therefore have a bearing on the faecal pH. The study found that diet 6 fed only 3 meals per day whereas diet 5, which fed a much greater percentage of concentrate, fed 4 meals per day and subsequently had a higher faecal pH. Although no apparent correlation existed between feed or faecal pH, the results of the experiment indicate that faecal pH may be related to the digestibility of the products within the diet and not feed pH *per se*.

References

Canh, T.T, Sutton, A.L., Aarnink, A.J.A., Verstegen, M.W.A., Schrama, J.W. and Bakker, G.C.M. (1998). Dietary carbohydrates alter the faecal composition and pH and the ammonia emission from slurry of growing pigs. *Journal of Animal Science* 76:1887-1895.

Kienzle, E. (1994). Small intestinal digestion of starch. *Revue Mediciene Vetinaire* 145(2):199-204.

B.M.L. McLean, J.J. Hyslop, A.C. Longland, D. Cuddeford and T. Hollands (2000). Physical effects of processing barley and its effect on intra-caecal fermentation parameters in ponies. *Animal Feed Science and Technology* 85:79-87.

Effects of dietary lipids on the erythrocyte morphology of ponies

C Horne, L Clark and R Geering
Department of Agricultural Sciences, Imperial College London (Wye Campus), Wye, Ashford, Kent, TN25 5AH, UK. Email, r.geering@imperial.ac.uk

Introduction

Many dietary oils have used in equine feeds primarily to improve joints and coat. Studies in rats have demonstrated the importance of dietary lipids in maintaining the normal discoid shape of the erythrocyte (Escadero *et al.*, 1998). It is not yet known whether the addition of lipids to the diet can change the physiology of equine erythrocytes. The aim of this study was to identify the effects of sunflower oil (SF), green-lipped mussel oil (GLM), linseed oil (LN) and soya oil (SY) on erythrocyte shape and tendency to form rouleauxs in the horse.

Materials and method

Two New Forest ponies of similar age and weight and on clinical screening were used in the study. They were fed either the control diet (C) of silage *ad libitum* or the control diet plus 50ml of either SF, GLM, LN or SY. Bloods were taken prior to the start of dietary manipulation and at the end of the two-week experimental period. There was a "wash-out" period of two weeks between each treatment on the control diet. Samples were processed as described by Berga *et al.* (1982) and viewed under a Hitachi S-430 Scanning Electron Microscope at an accelerated voltage of 15kV and at a magnitude of 3K. The number of discocytes, stomatocytes, echinocyte 1,2 and 3, and rouleaux formation were counted. Statistical differences between the groups were assessed using the Analysis of Variance.

Results

Supplementing the diet with SF increased the number of discocytes compared to the GLM ($P<0.05$) and C (no statistical significance) groups, whilst both SF and GLM caused a reduction ($P<0.05$) in the number of stomatocytes observed but no differences with respect to echinocytes (Table 1). There was also a greater ($P<0.05$) formation of rouleaux in the GLM groups. Surprisingly there was no difference in the distribution of discocytes, stomatocytes, total echinocytes, and echinocyte type or rouleaux formation with the addition of LN and SY to the diet.

Table 1.
Effect of dietary lipids on erythrocyte morphology.

	Control	SF	GLM
Discocytes (%)	52±11	78±2	40±6
Stomatocytes (%)	20±3	8±8	11±3
Rouleaux (%)	14±7	6±2	44±8

Conclusion

The results demonstrate that dietary lipids have a pronounced impact on erythrocyte morphology and ability to aggregate. The beneficial effect of SF is likely to partly be due to its high a-linoleic acid content, which is essential in preventing echinocyte formation. GLM has probably promoted aggregation by increasing fluidity of the membrane due to incorporation of long chain polyunsaturated fatty acids in to the phospholipids. Further work is required on the relationship between dietary lipids, erythrocyte morphology and the athletic performance of the horse.

References

Berga L, Vives-Corrons JL, Ferran MJ, Rozman C, Aguilar I Bascompte JL, Jou J, Feliu E, Ester A, Dolz J. (1982). Contributions of scanning electron microscopy (SEM) to the study of erythrocyte morphology and to the diagnosis of hemolytic syndromes. *Sangre (Barc)*. **27**(4A): 445-60. Spanish only.

Escudero A, Montilla JC, Garcia JM, Sanchez-Quevedo MC, Periago JL, Hortelano P, Suarez MD. (1998). Effects of dietary (n-9), (n-6), and (n-3) fatty acids on membrane lipid composition and morphology of rat erythrocytes. *Biochemica et Biophysica Acta* 1394: 65-73.

Inhibition of equine cyathostomes to glucosinolate hydrolysis products

L Brett, J Rossiter, L Clarke and R Geering
Department of Agricultural Sciences, Imperial College London (Wye Campus), Wye, Ashford, Kent, TN25 5AH, UK, Email, r.geering@imperial.ac.uk

Introduction

Cyathostome nematodes have become the central focus of concerns related to internal parasites of horses due to their pathological and persistent effects on the growing and mature horse (Murphy and Love, 1997). Current anthelmintic treatments have, however, led to resistance and alternative control measures are required. Glucosinolate hydrolysis derivatives have a wide range of biological activities, and have previously been shown to inhibit root nematodes (Brown and Morra, 1997). The aim of the study was to examine whether cyathostome nematode eggs could be inhibited by glucosinolate hydrolysis products.

Materials and methods

An *in vitro* egg hatch assay was set up using eggs extracted from the faeces of naturally infected horses (Kassai, 1999). The egg suspension (containing approximately 100 eggs) were incubated with nine different concentrations (range: 20-10,000 ppm) of the glucosinolate hydrolysis products (allyl isothiocyanate (C), allyl nitrile (N) and phenethyl isothiocyanate (P). A control was set up in which water replaced the hydrolysis product. The number of eggs and larvae were counted using an inverted microscope. Statistical differences between treatments were assessed using General Linear Model, ANOVA.

Results

P and C inhibited egg hatching to L1 larvae at concentrations of 50 and 2000 ppm, respectively but N appeared to have no influence on egg hatching within the experimental concentration range (Table 1).

Conclusions

Hatching of eggs to L1 larvae was inhibited by both the isothiocyanates but not the nitrile. The difference in their responses may, in part, be due to the variation in their structure. The data also suggests that aromatic compounds are more toxic than aliphatic ones. Further work is required to assess their suitability for commercial application.

Table 1. Effect of allyl isothiocyanate (C) allyl nitrile (N) and phenenthyl isothiocyanate (P) on % egg hatch.

Concentration (ppm)	Control (%)	C (%)	N (%)	P (%)
20	79	86	92	65
50	74	89	90	35
100	75	88	81	25
200	76	76	80	12
500	81	81	79	5
1000	83	65	73	2
2000	83	21	87	3
5000	77	8	82	0
10000	79	0	83	0

References

Brown, P.D. and Morra, M.J. (1997). Control of soil borne pests using glucosinolate containing plants. *Advances in Agronomy* 61: 167-231.

Kassai, T. (1999). Nematoa, horses and diagnostic helminthology. *Veterinary Helminthology,* 169: 183-203.

Murphy, D. and Love, S. (1997). The pathogenic effects of experimental cyathostome infections in ponies. *Veterinary Parasitology,* 106(4): 99-110.

Nutrient digestion and faecal excretion of three coarsely mixed diets in ponies

A S Chaudhry
School of Agriculture, Food and Rural Development, University of Newcastle, Newcastle-upon-Tyne NE1 7RU, UK
E-mail: a.s.chaudhry@ncl.ac.uk

Introduction

Adequate feeding of a horse is essential to help sustain its health and performance in different situations. Traditionally, horses are fed hay or grazed grass without or with grains plus supplements or compound feeds. However, the practice of feeding horses varies greatly in the UK depending upon the workload of a horse, co-location of the owner and the seasonal variations mainly due to the weather and the availability of pasture or hay. Although many complete feeds are also available commercially, their nutritive values do vary and thus they may not suit a horse. Therefore, there is a need to formulate nutritious diets that are palatable for a horse and cost-effective for the owner. It is also imperative to prepare diets that are more completely digested to minimise faecal contamination the environment. This study compared the suitability of three coarsely mixed diets to improve digestion and reduce excretion in adult ponies.

Materials and methods

The diets contained either 50% dried Lucerne (LN, diet A), or 23% LN plus 26% straw and 10% hay (diet B), or 25% LN plus 20% brewers' grains (BG, diet C), as the main sources of neutral detergent fibre (NDF, fibre). Appropriate amounts of oil, processed barley, maize, wheat, peas and beans were also present in these diets. The diets were iso-nitrogenous but different in other nutrients (Table 1). Each diet was fed to three aged ponies with initial mean ($\pm$SE) live-weight (LW) of 329$\pm$39kg over three periods according to a 3x3 Latin square design. The ponies were housed individually on wood shavings at the Bristol University's Langford Veterinary School. About 5kg of each diet was fed daily in two equal portions at 8:30am and 4:30pm to the relevant horse for 21 days in each period. Fresh drinking water was always available to these ponies. After 14 days of adaptation to housing and diets, the total faeces from each horse were collected at 8 hourly intervals for up to 7 days. The faeces were dried at 70°C and weighed to estimate faecal dry matter (DM) output at each collection time. The dried faeces for each diet in each period were then pooled and analysed together with samples of each diet to estimate nutrient composition and digestibility of each diet for each pony. The data were statistically analysed to examine if the effect of diet on digestibility was significant (Sign) at $P<0.05$. The data on faecal output, averaged over ponies, were also plotted in Excel to illustrate the effect of diets on the patterns of faecal output over various times of faecal collection post first feeding of each diet.

Table 1.
Composition, intake (*kg*) and digestibility of diets.

Diets	A	B	C	SE
Composition (g/kg DM)				
Organic matter, OM	931	928	946	ND
Crude Protein, CP	151	158	146	ND
Neutral detergent fibre, NDF	278	368	340	ND
Gross energy, GE, MJ	18	18	19.4	ND
Food intake, DM/pony/day	4.7	4.7	4.3	ND
Digestibility (g/kg)				
DM (DMD)	733	661	727	15*
CP (CPD)	842	840	851	9ns
NDF (NDFD)	335	328	444	78ns
GE (GED)	711	642	725	14*
DE, estimated MJ in vivo	12.8	11.6	14.1	0.06**

ND, not determined; se, standard error; *, $P<0.05$; **, $P<0.01$

Figure 1.
Faecal output of different diets.

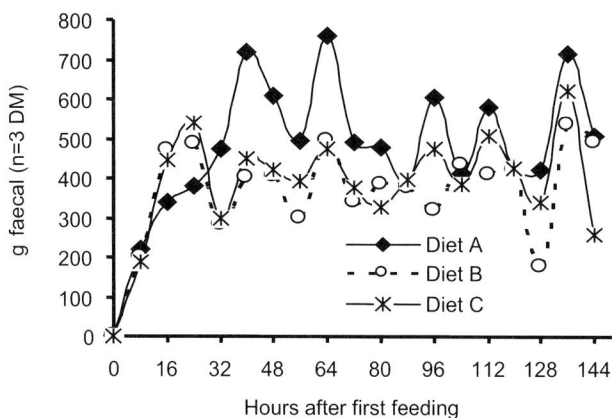

Results

Table 1 showed that diet B contained higher NDF but lower GE which was due to the presence of more fibrous straw plus hay than diet A and C which contained relatively less fibrous LN and BG respectively. The diets differed significantly for DMD, GED and DE (P<0.05) but not for CPD and NDFD (P>0.05). While diets A and C showed significantly higher DMD, GED (P<0.05) and DE (P<0.01) than diet B, diet C had significantly greater DE than diet A (P<0.01). Also, diet C tended to show greater NDFD than diets A and B which were about 35 and 32% more than diet B and A respectively. While the shapes of curves for faecal outputs of different diets were almost comparable (Figure 1), the rates and extents of faecal output varied with time. The faecal output for diets A and C peaked at 24 hours whereas for diet B, it peaked between 40 and 64 hours post first feeding. Diet B yielded more faeces than diets A and C at almost all times which confirmed that the diet B was least digestible and thus most wasteful. This was because the high fibrous contents of this diet were less utilised by the horse and thus should be less preferred as major constituents of a horse diet. However, it would be desirable to optimise the level and type of fibrous feed ingredients which may otherwise benefit the horse gut and behaviour, help reduce faecal excretion and thus the cost of feeding a horse.

Conclusion

It was possible to reduce the faecal output and thus improve nutrient digestion by modifying the ingredient composition of a horse diet. Clearly the diets were palatable, nutritious and the amounts that were consumed daily by these ponies were sufficient to help maintain their apparent health. In fact, all ponies were able to gain similar but modest body weights confirming once again that although the diets varied in their utilisation by the ponies, they were nutritionally balanced for the ponies of this study.

Acknowledgements

To C. Fintl, P. Liles and many staff for their help during this study at the University of Bristol, UK.

Equine feeding side preference – incidence and age effect

S H Bottom[1], H Owen[1], R E Lawson[1], P A Harris[2] and S Hall[2]
[1]School of Land-based Studies, The Nottingham Trent University, Brackenhurst Nottingham Road, Southwell, Nottinghamshire NG25 0QF, UK, Email: sarah.bottom@ntu.ac.uk
[2]Equine Studies Group, WALTHAM Centre for Pet Nutrition , Melton Mowbray, Leicestershire LE14 4RT, UK

Introduction

Two choice preference testing is used in animals to determine food preference or taste response (Nicol, 1997). This relies upon selection from a theoretical mean of 0.5, if no preference is shown. The aims of this study were two fold: to measure the incidence of side preference in the horse when given the choice of two identical feeds; and to assess the influence of age on side preference.

Materials and methods

18 TB x horses were used in this study. Six horses were allocated to each of the age groups; young (2-4 years), middle (8-14 years) and old (over 16 years). All horses were housed in stables of identical design. 1.5kg DM of basal feed was offered in two containers, positioned left and right, for a ten-minute period over nine days. All spillages were collected and recorded. Intake was calculated as the (amount of feed offered – amount of feed remaining – 0.5x the spillage). Individual ingestive behaviours were recorded during the presentation period using individual CCTV cameras and video recorders. Eat frequency was recorded as the number of times the horse ate from a container, relevant to the side selected. A two-sample t–test was used to analyse individual and age group left / right intake. Interaction between left/right intake and age groups was analysed using two-way ANOVA. Chi squared was applied to the behaviour left / right eat frequency to determine differences between age groups and individual.

Results

The influence of age on side preference is shown in Figure 1. An interaction between age and left/right preference was found (P<0.01), with the strongest response being observed in the old group (P<0.01).

This was the result of three horses showing left preference in the old group. The old group had the highest total eat frequency, supported by the highest total intake ($P<0.001$). Increased non-ingestive behaviours such as walk resulted in the lowest eat frequency in the young group and reduced total intake.

Figure 1.
Mean age group
side intake

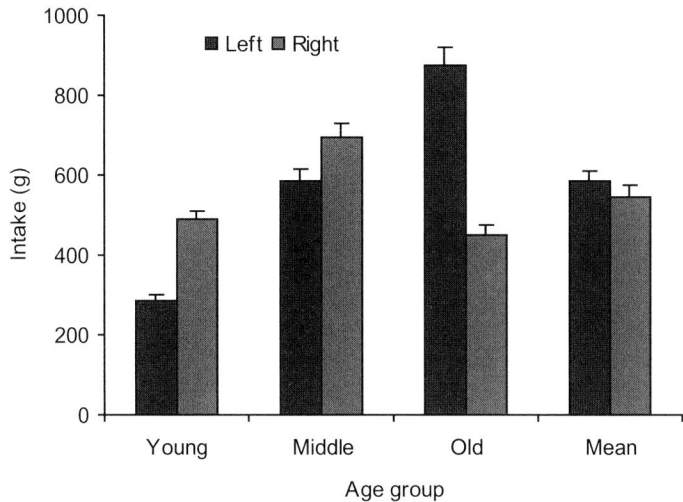

Figure 1. Mean age group side intake

Age group	Mean left intake (g)	s.e.m.	Mean right intake (g)	s.e.m.	Side preference
Young	502	42.2	777*	99.2	right
Young	319	166	1161*	168	right
Middle	944*	71.3	681	68.6	left
Middle	550	172	1066**	188	right
Old	475	96.3	912*	88.2	right
Old	1254***	41	222	40.5	left
Old	982*	168	319	163	left
Old	1164***	87.8	95	55.6	left

Table 1.
Individuals
showing side
preference
(difference
between left/
right intake
*p<0.05,
**p<0.01,
***p<0.001).

Individual horses showing side preference are given in Table 1, with a side preference being recorded in eight of the 18 horses studied. A significant left preference recorded in two horses contradicted a significant difference in left *versus* right eat frequency. Limited sampling behaviour by the horses, followed by consistent intake from the preferred container, resulted in a low eat frequency.

Conclusion

Horses, when offered a choice of two feed containers may consistently select more from one side. This is proposed to be awareness of no benefit in selecting between the two choices. This may be more evident in older horses due to learning associated with postingestive consequences and reduced exploratory behaviour. Side preference has implications in the use of preference testing when there is no cost in selecting one choice continuously.

Reference

Nicol, C.J., (1997). Environmental choice of farm animals. *Animal Choices* 20: 35-43.

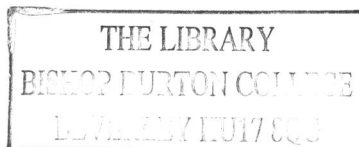

THE LIBRARY
BISHOP BURTON COLLEGE
LIBRARY HU17 8QG

Relationship between crude protein intake and water intake in forage based equine diets

J J Hyslop
University of Edinburgh, Dept Vet Clinical Studies, Easter Bush, Roslin, Midlothian EH25 9RG, UK
(Current address: ADAS Redesdale, Rochester, Otterburn, Newcastle upon Tyne, NE19 1SB. UK, E-mail: jimmy.hyslop@adas.co.uk)

Introduction

As a result of the need to excrete the waste products of protein catabolism in urine, increases in protein intake have been associated with increases in water intake in poultry (Bailey, 1990), in pigs (Brooks and Carpenter, 1990) and in ruminants (Agricultural Research Council, 1980). Whilst there is only very limited data available to support a similar relationship in horses (Meyer, 1987) it is believed that increases in protein intake in equines also result in greater water consumption when high protein diets are fed (Lawrence, 1998). The objective of the current study was to examine the relationship between crude protein intake and water consumption when forage based diets were offered *ad libitum* to individually fed ponies.

Materials and methods

Six mature Welsh-cross pony geldings with a mean liveweight (LW) of 281 kg (s.e.d. 0.89) were individually housed in pens bedded with wood shavings and used in a changeover design experiment consisting of two 21-day periods. Ponies were given *ad libitum* access to a metered water supply and either short-chop de-hydrated grass (DHG) or traditional grass hay (HAY) which had been harvested from the same grass crop. 60 g/pony/day of a mineral and vitamin supplement was also given once daily. The grass crop was harvested from a mixed perennial ryegrass/timothy sward in early August and half the crop was immediately chopped, de-hydrated in a low temperature drier (110 – 140°C) and vacuum packed as for commercial distribution. The other half was left in the field and made into traditional grass hay by mechanical tedding and sun curing for a period of 6 days, after which it was baled conventionally. Each 21-day period consisted of a 16-day adaptation phase and a 5-day recording phase when voluntary dry matter intake (DMI), crude protein intake (CPI), water consumption and *in vivo* apparent digestibilities were determined by total faecal collection. During the 5-day collection phase, wood shavings were replaced by rubber mats to facilitate faecal collection and water intakes were recorded for each pony using the individually metered water supply. Composite samples of feed and faeces were analysed for organic matter (OM), crude protein (CP), acid detergent fibre (ADF), neutral detergent fibre (NDF) and gross energy (GE) contents by routine laboratory methods (AOAC, 1990). The water holding capacity (WHC) of both forages was also determined using a modification of the centrifugation method described by Robertson and Eastwood (1981) where WHC is expressed as millilitres of water per gram of DM (ml/ g DM). Triplicate, 0.5g samples of each forage were placed in a 25ml centrifuge tube with 20 ml of distilled water and allowed to soak for 24 h. Each tube was then centrifuged at 5000 rpm for 20 minutes, the supernatant allowed to drain off for 1 h and the fresh weight of feedstuff determined. The sample was then oven dried at 60 °C for 48 h and WHC calculated. Statistical analysis of intake parameters was carried out by analysis of variance and the water intakes of individual ponies related to a range of nutrient intake parameters by simple linear regression techniques using Genstat 5 (Lawes Agricultural Trust, 1993).

Results

Detailed DMI and *in vivo* apparent digestibility results have been presented previously (Hyslop *et al*, 1998) and only results relating to water consumption will be discussed here. The chemical composition and WHC's of both DHG and HAY along with average pony DMI, CPI and water intake figures are given in Table 1.

Table 1. Chemical composition of feeds (g/kg DM unless otherwise stated), WHC, DMI, crude protein intake (CPI) and water intake (WI) in ponies given either DHG or HAY *ad libitum*.

Chemical composition			Intake parameter and WHC				
	DHG	HAY		DHG	HAY	s.e.d	Sig
DM g/kg	908	833	DMI (kg/d)	7.61	7.70	0.218	NS
OM	929	955	DMI (g/kg LW)	26.7	27.8	0.720	NS
CP	172	85	DMI (g/kg $LW^{0.75}$)	110	113	2.940	NS
ADF	300	332	Crude protein	1314	661	82.6	**
NDF	677	695	intake (g/d)				
GE MJ/kg DM	19.3	19.9	Water intake (l/d)	30.5	25.2	0.520	***
			Water intake (l/kg DMI)	4.11	3.32	0.184	*
			WHC (ml/g DM)	10.5	11.2	0.346	NS

* $P<0.05$, ** $P<0.01$, *** $P<0.001$.

In comparison to HAY, the DHG had higher DM and CP contents but lower OM, ADF, NDF and GE contents. Although the WHC was slightly higher for the HAY compared to the DHG, this difference did not reach statistical significance. DMI averaged 7.66 kg/d and did not differ significantly between the two forages when compared on a daily (kg/d), pony LW (g/kg LW) or pony metabolic LW (g/kg $LW^{0.75}$) basis. At 1314 g/d, CPI was over twice as high when DHG was fed compared to 661 g/d when HAY was offered. Average water consumption at 30.5 l/d was significantly higher (P<0.001) on the DHG forage compared to the figure of 25.2 l/d when HAY was offered. When related to daily DMI, water intake remained significantly higher at 4.41 l/kg DMI when DHG was fed compared to 3.32 l/kg DMI when HAY was offered. This suggests that some attribute of DHG led to increased water consumption compared to HAY.

Regression analysis of individual pony values showed that most dietary variables were poorly correlated with water intakes (data not shown). However, there was a significant linear relationship (Figure 1) between water intake (WI) and crude protein intake (CPI) where water intake is expressed in l/day and CPI is expressed in g/day.

Figure 1. Relationship between crude protein intake and water intake in ponies given one of two forages *ad libitum*.

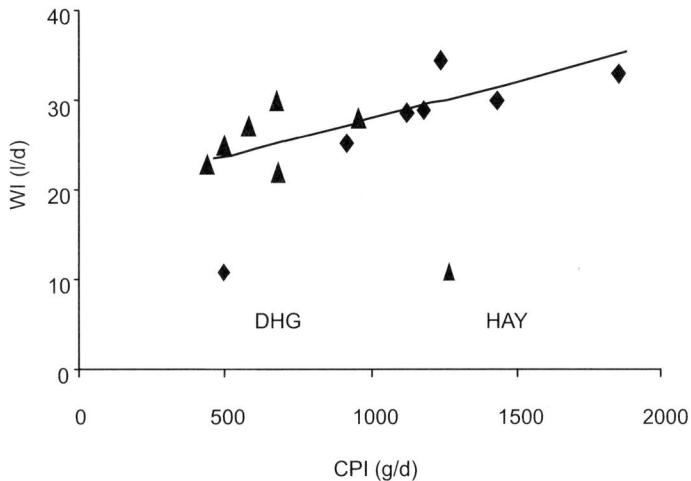

Linear regression relationship:
WI = 19.61 (s.e. 2.48) + 0.00838 (s.e. 0.00233) CPI
(R^2 0.521: RSD 3.26: P<0.01)

Conclusions

Given that WHC did not differ significantly between DHG and HAY and in fact was lower for the DHG forage, it is unlikely that the increased water intake when DHG was fed can be related to any inherent association between water and the fibre structure of DHG compared to HAY. Although carried out with only two forage based diets, the linear relationship between crude protein intake and voluntary water consumption observed in this study does provide tentative evidence to support the hypothesis that increased water consumption is associated with higher protein intakes. However, more research is clearly needed in this area to examine the factors that may interact to influence water consumption over a wide range of dietary and environmental conditions. In practice, horse owners feeding high protein diets should be urged to ensure that adequate water is freely available to all animals and be aware that bedding requirements should be sufficient to accommodate high urine output.

Acknowledgements

This work was partly funded by the Horserace Betting Levy Board.

References

Agricultural Research Council. (1980). *The Nutrient Requirements of Ruminant Livestock*. CAB International, Wallingford, UK.

Association of Official Analytical Chemists (AOAC). (1990). *Official Methods of Analysis of the AOAC*. 15th Edition. Virginia, USA.

Bailey, M. (1990). The water requirements of poultry. In: *Recent Advances in Animal Nutrition*. Edited by Haresign, W. and Cole, D.J.A., Butterworths, London. UK, pp. 161-176.

Brooks, P. H. and Carpenter, J. L. 1990. The water requirements of growing and finishing pigs – theoretical and practical considerations. In: *Recent Advances in Animal Nutrition*. Edited by Haresign, W. and Cole, D.J.A., Butterworths, London. UK, pp. 115-136.

Hyslop, J. J., Bayley, A., Tomlinson, A. L. and Cuddeford, D. (1998). Voluntary feed intake and apparent digestibility *in vivo* in ponies given *ad libitum* access to de-hydrated grass or hay harvested from the same grass crop. *Proceedings of the British Society of Animal Science*, BSAS, Penicuik, UK, p. 131.

Lawes Agricultural Trust. (1993). *Genstat 5 Reference Manual*. Rothemstead Experimental Station, Harpenden, Hertfordshire, UK.

Lawrence, L. M. (1998). Protein requirements of equine athletes. In: *Advances in Equine Nutrition*. Edited by Pagan, J.D., Nottingham University Press, Nottingham, UK, pp. 161-166.

Meyer, H. (1987). Nutrition and the Equine Athlete. In: *Equine Exercise Physiology 2*. Edited by Gillespie, J.R. and Robinson, N.E., ICEEP Publications, Davis, California, USA, pp. 644-673.

Robertson, J. A. and Eastwood, M. A. (1981). A method to measure the water holding properties of dietary fibre using suction pressure. *British Journal of Nutrition* 46: 247-255.

Pagan, J. D. (1998). Protein requirements and digestibility: a review. In: *Advances in Equine Nutrition*. Edited by Pagan J.D., Nottingham University Press, Nottingham, UK, pp. 43-50.

Relationship between water holding capacity and fibre concentration in equine feeds

J J Hyslop[1*], B M L McLean[1], and M J S Moore-Colyer[2]
[1]University of Edinburgh, Dept Vet Clinical Studies, Easter Bush, Roslin, Midlothian EH25 9RG, UK (*Current address: ADAS Redesdale, Rochester, Otterburn, Newcastle upon Tyne, NE19 1SB, UK, E-mail: jimmy.hyslop@adas.co.uk)
[2]Institute of Rural Studies, University of Wales Aberystwyth, Llanbadarn Fawr, Aberystwyth SY23 3AL, UK

Introduction

The water content of feedstuffs is an important factor when considering both the nutritive value of a feed for dietary rationing purposes and when assessing the functional properties associated with the practical inclusion of a feedstuff in animal diets. Water is a vital nutrient in its own right and must be supplied on a daily basis. Feed associated water provides one source of this obligatory requirement. In addition, the water holding capacity (WHC) of a feedstuff and its relationship with other constituents of the feed may have important effects. Water can be associated with feedstuff in one of three ways (Robertson and Eastwood, 1981b). Firstly, water can be bound by the hydrophilic polysaccharides of the fibre component of feeds. Secondly, water can be held within the structural fibre matrix of feeds and finally, water can be associated with feedstuff fibre other than bound or matrix water and is usually considered as water trapped within the cell wall lumen. Feed associated parameters such as digesta passage rate, digesta hydration rates following consumption, which in turn may effect rate of digestion and voluntary feed intake levels could all be influenced by the relationship between WHC and the dry matter constituents of a particular feed. WHC has been suggested as a contributor to the regulation of voluntary feed intake (VFI) in both ruminants (Froetschel and Amos, 1991) and pigs (Tsaras, Kyriazakis, and Emmans, 1998) as a result of the bulky or fibrous nature of high WHC feeds. However, as yet there are no published data-sets where the WHC of a wide range of feed types has been related to either the fibre or protein content of those same feeds. The objective of this study was to examine the relationship between WHC and the dry matter (DM), organic matter (OM), crude protein (CP), acid detergent fibre (ADF), neutral detergent fibre (NDF) and cell wall (CW) contents across a wide range of equine feedstuffs of both concentrate and forage types.

Materials and methods

During the period 1995 – 2000 a total of 54 feedstuffs were examined in a series of equine nutrition studies at either Edinburgh University or the Institute of Rural Studies. These feedstuffs can be loosely grouped into one of six categories as follows:

(1) rolled, micronised or extruded barley, maize, peas, wheat and naked oats (CEREALS)
(2) commercially available equine concentrate feeds (CONC)
(3) fibre based straight feeds or pelleted complete diets (FIBRE)
(4) alfalfa based dried forages (ALFALFA)
(5) grassland based preserved forages including chopped silage, baled silage, haylage, hay and de-hydrated grass (GRASS)
(6) cereal straws from oats, barley and wheat (STRAWS).

Composite samples of all 54 feedstuffs were analysed for DM, OM, CP, ADF, and NDF contents by routine laboratory methods (AOAC, 1990). CW content for all feeds was defined as NDF-ADF contents. The water holding capacity (WHC) of all 54 feedstuffs was also determined using a modification of the centrifugation method described by Robertson and Eastwood (1981a) where WHC is expressed as millilitres of water per gram of DM (ml/g DM). Triplicate, 0.5g samples of each feedstuff were placed in a 25ml centrifuge tube with 20 ml of distilled water and allowed to soak for 24 h. Each tube was then centrifuged at 5000 rpm for 20 minutes, the supernatant allowed to drain off for 1 h and the fresh weight of feedstuff determined. The sample was then oven dried at 60°C for 48 h and WHC calculated. Simple descriptive statistics for each of the 6 categories were calculated. Correlation co-efficients (r) between feed compositional variables and WHC for the data-set of all 54 feedstuffs were estimated and simple linear and quadratic relationships determined using Genstat 5 (Lawes Agricultural Trust, 1993).

Results

The correlation matrix amongst feedstuff composition and WHC is given in Table 1. Fibre constituents (ADF, NDF and CW) were all positively correlated with WHC (r ranged between 0.72 – 0.88). Whilst CP was itself negatively correlated with WHC (r = -0.54), its relatively high auto-correlations with fibre components (r ranged between – 0.51 - -0.57) meant that it was unsuitable as a predictor variable in multiple regression equations with any of the fibre components. The best predictor of WHC across all 54 feedstuffs was the quadratic relationship with NDF content (Figure 1). The prediction equation is given below. Descriptive statistics for each of the six feed categories are given in Table 2. Average WHC (ml/g DM) across the six feed categories ranged from 2.9 for CEREALS to 13.9 for STRAWS whilst average NDF contents ranged from 119 – 875 g/kg DM across the same two categories.

Table 1. Correlation matrix between feed composition and WHC in 54 equine feedstuffs.

WHC	1.00						
DM	-0.41	1.00					
OM	-0.24	0.23	1.00				
CP	-0.54	0.07	0.09	1.00			
ADF	0.87	-0.24	-0.44	-0.54	1.00		
NDF	0.88	-0.22	-0.45	-0.57	0.96	1.00	
CW	0.72	-0.16	-0.38	-0.51	0.68	0.86	1.00
	WHC	DM	OM	CP	ADF	NDF	CW

WHC prediction equation:
WHC = $2.259 + 0.00261 \times NDF + 0.00001316 \times NDF^2$
($R^2 = 0.81$; RSD = 1.71; P<0.001).

Figure 1. Quadratic relationship between WHC (ml/g DM) and NDF content (g/kg DM) in 54 equine feedstuffs.

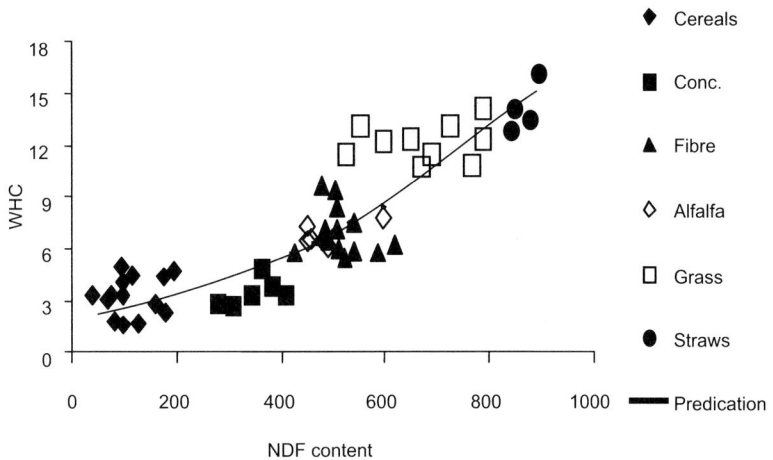

Conclusions

The water holding capacity of equine feedstuffs is highly correlated with the fibre components of the feed dry matter and can be accurately predicted in practice using a simple quadratic relationship with NDF content.

Acknowledgements

The Horserace Betting Levy Board and Dodson & Horrell Ltd for funding (B.M.L. McL).

Table 2. Descriptive statistics of feed composition and WHC for the six feed categories used in this data-set.

Category	Parameter	(g/kg) DM	OM	(g/kg DM) CP	ADF	NDF	CW	(ml/g DM) WHC
CEREALS (n = 15)	Mean	899	980	168	38	119	81	2.9
	Range	884-924	971-990	86-368	15-67	44-201	28-137	1.4-4.7
	s.d.	9.9	6.4	93.4	19.8	45.1	33.4	1.17
	% C.V.	1.1	0.6	55.4	52.5	38.1	41.3	39.5
CONC (n = 6)	Mean	890	940	130	142	354	212	3.3
	Range	865-909	912-957	98-157	94-194	286-415	186-230	2.4-4.6
	s.d.	18.2	17.4	25.0	39.1	48.3	15.3	0.8
	% C.V.	2.0	1.6	19.2	27.6	13.6	7.2	24.5
FIBRE (n = 14)	Mean	878	912	118	283	520	237	6.7
	Range	832-933	890-924	78-158	225-438	430-623	153-276	5.3-9.5
	s.d.	30.8	8.2	23.1	54.5	47.5	33.5	1.4
	% C.V.	3.5	0.9	19.7	19.3	9.1	14.1	20.0
ALFALFA (n = 5)	Mean	893	912	134	361	496	134	6.6
	Range	860-937	906-925	98-161	228-436	457-602	71-236	5.9-7.5
	s.d.	34.5	7.9	23.8	84.7	61.9	69.1	0.7
	% C.V.	3.9	0.9	17.8	23.4	12.5	51.5	10.0
GRASS (n = 10)	Mean	744	945	89	397	681	284	11.9
	Range	337-935	918-982	44-172	300-467	529-795	119-377	10.5-13.8
	s.d.	21.0	22.0	43.7	58.5	95.7	80.8	1.1
	% C.V.	28.2	2.3	49.3	14.7	14.1	28.4	8.9
STRAWS (n = 4)	Mean	895	957	26	555	872	317	13.9
	Range	831-930	936-984	19-36	541-565	849-898	299-333	12.7-15.9
	s.d.	44.9	20.4	7.5	11.6	23.4	14.1	1.4
	% C.V.	5.0	2.1	28.6	2.1	2.7	4.5	10.0

s.d. = standard deviation; C.V. = co-efficient of variation.

236

References

Association of Official Analytical Chemists (AOAC). (1990). *Official Methods of Analysis of the AOAC*. 15th Edition. Virginia, USA.

Froetschel, M.A. and Amos, H.E. (1991). Effects of dietary fibre and feeding frequency on ruminal fermentation, digesta water holding capacity and fractional turnover of contents. *Journal of Dairy Science* **69**: 1312-1321.

Lawes Agricultural Trust. (1993). *Genstat 5 Reference Manual*. Rothemstead Experimental Station, Harpenden, Hertfordshire, UK.

Robertson, J. A. and Eastwood, M. A. (1981a). A method to measure the water holding properties of dietary fibre using suction pressure. *British Journal of Nutrition* **46**: 247-255.

Robertson, J. A. and Eastwood, M. A. (1981b). An investigation of the experimental conditions which could affect water holding capacity of dietary fibre. *Journal of the Science of Food and Agriculture* **32**: 819-825.

Tsaras, L.N., Kyriazakis, I. and Emmans, G.C. (1998). The prediction of the voluntary food intake of pigs on poor quality foods. *Animal Science* **66**: 713-723.

The effect of particle size on volatile fatty acid profiles obtained from an alfalfa and unmolassed sugar beet pulp diet following *in vitro* incubation with equine faeces

RKT Bice and MJS Moore-Colyer
Institute of Rural Studies, University of Wales Aberystwyth, Llanbardan Fawr, Aberystywth, SY23 3AL, UK

Introduction

The equid has been estimated to derive 70% of energy from volatile fatty acids (VFA) production. Hind gut fermentation in the equid degrades structural carbohydrates to VFA's that are oxidised by the liver as a source of energy. Maximising this potential source of energy is an important consideration when attempting to improve the utilisation of fibre feeds by the equid. It is generally accepted that the acetate:propionate:butyrate are produced in a ratio of 76:18:5 by horses consuming forage diets (Moore-Colyer 2000). This optimum can be unbalanced by the inclusion of high levels of cereal-based concentrates in the diet. Therefore to maximise hindgut function it is important to maximise the use of high-energy fibrous feeds such as alfalfa and sugar beet pulp (USBP), while reducing the inclusion of concentrate feeds. This experiment sought to determine the effect of feed preparation, namely particle size on VFA profiles from a range of alfalfa : unmolassed sugar beet pulp diets.

Materials and methods

Following the *in vitro* degradation of alfalfa and USBP using the gas production technique of Theodorou *et al.* (1994), VFA production profiles were determined from culture fluid samples, employing the gas chromatography method. The feeds used were chopped (C) and ground (G) alfalfa (CAA, GAA) and unmolassed sugar beet pulp (CSBP, GSBP) and in the following ratios of alfalfa and USBP 90:10 (C10, G10), 80:20 (C20, G20), 70:30 (C30, G30). Two replicates of each residue were acidified using H_3PO_4 and stored at 4°C for 12hr, to enable the precipitation of proteinaceous material. Samples were then centrifuged and internal and external standards added to the supernatant. A Chrompack 9000 gas chromatograph equipped with an automatic liquid sampler and split injection and flame ionisation detector was used to analyse the samples. Analysis of

Table 1. Proportions of the main volatile fatty acids produced following the incubation of ten alfalfa and USBP food combinations of with a faecal inoculum from ponies fed the corresponding diets.

	CAA	CSBP	C10	C20	C30	GAA	GSBP	G10	G20	G30	s.e.d
Acetate	64.99[e]	61.56[b]	63.15[b]	64.56[de]	64.01[cd]	63.61[bc]	59.71[a]	61.66[b]	61.54[b]	62.21[b]	0.269
Propionate	27.12[a]	31.37[c]	29.77[b]	27.56[a]	28.53[a]	28.39[a]	32.44[d]	31.57[c]	30.79[c]	31.06[c]	0.856
Iso Buty + N Buty	7.89[d]	7.07[ab]	7.08[b]	7.87[d]	7.46[bc]	7.99[d]	7.85[d]	6.77[a]	7.67[cd]	6.73[a]	0.136

[abcde] Values in the same row not sharing common superscripts differ significantly (P<0.01).

variance (Genstat 5) was used to determine if significant differences existed between VFA profiles obtained from the different feed combinations.

Results

Table 1 illustrates the proportions of the three main volatile fatty acids for each treatment. CAA shows a significantly higher proportion of acetate than all other samples. The CSBP and the GSBP produced significantly less acetate than the alfalfa samples in both chopped and ground form respectively. Propionate values differed significantly (P<0.01) between the chopped and ground substrates. CSBP was significantly less than GSBP. C10 was significantly different from the other chopped samples and G10 showed a greater proportion of propionate compared to G20 and G30, although this was not significant. Proportions of butyrate were significantly larger for C20 and G20 compared with the values for 10% and 30% SBP inclusion for the respective treatments. CSBP and GSBP had significantly different VFA profiles, whereas CAA and GAA showed similar levels of both butyrate and propionate.

Conclusion

The results obtained from this experiment indicate that alfalfa and unmolassed sugar beet pulp are both readily degraded feeds when exposed to equid hindgut microorganisms. However, these results also show that VFA profiles are influenced by both diet composition and feed particle size, with the more readily degraded ground feed producing significantly higher propionate levels than the more slowly degraded chopped feeds. This indicates that to maintain a healthy VFA profile within the hindgut, forage feeds should be fed in a chopped form.

References

Moore-Colyer M.J.S. (2000) Studies on the degradation kinetics of botanically diverse fibrous feeds and their apparent digestibility and rate of passage when fed to ponies. *PhD Thesis*. University of Edinburgh.

Theodorou, M.K., Williams, B.A., Dhanoa, M.S., and France J. (1994) A simple gas production method using a pressure transducer to determine the fermentation kinetics of ruminant feeds. *Animal Feed Science and Technology*, **48**, 185-197.

A preliminary investigation into the effects of forage type on ammonia levels in stables for horses

C J Blaken and R E Lawson
School of Land-based Studies, Nottingham Trent University, Brackenhurst, Nottingham Road, Southwell, Nottinghamshire, NG25 0QF, UK. Email: ruth.lawson@ntu.ac.uk

Introduction

Ammonia is generally present within all livestock buildings; it is a by-product of protein digestion. Chronic obstructive pulmonary disease (COPD) is probably the most common cause of chronic coughing in horses in the UK. This has resulted in horse owners offering haylage, which is a low-dust alternative to hay (Moore-Colyer and Longland, 2000). However, Moore-Colyer and Longland (2000) reported that haylage usually contains more protein than hay (70 *vs.* 44 g CP/kg). When protein supplied in the diet exceeds the requirements of the horse, urea production will increase and consequently ammonia output will increase. Aerial ammonia concentrations can be regarded as a predisposing factor of COPD (Pratt, 2000). The aim of this experiment was to investigate the effects of offering horses different basal forages (hay *vs.* haylage) on the aerial ammonia concentrations within stables at different times during the day.

Materials and methods

A total of 16 crossbred horses were used for this experiment. All horses were screened to eliminate horses with nasal discharge and coughing. The experiment was of randomised block design with animals balanced for sex, age and level of work; and housed in individual loose boxes. The stables were bedded with wood shavings; all faeces were removed on an hourly basis from 07:00 to 17:00, the areas of soiled bedding due to urination were left. Eight out of the 16 horses were offered meadow hay and the remaining eight horses were provided with haylage as the sole forage. The horses were offered their respective forages for at least one month prior to the start of the experiment. All horses were offered concentrate foods according to the recommendations of National Research Council, NRC (1989). Both the DP concentration of the forage (Cuddeford *et al.*, 1992) and the DP content of the concentrates were estimated (Glade 1984). Aerial ammonia concentrations were determined at

09:00, 13:00 and 17:00 using ammonia diffusion tubes (RAE systems) and forage samples were analysed using near infrared spectrometry. Mean ammonia recordings were subjected to the Mann-Whitney test for the two forage treatments and individual variations in the concentration of ammonia produced were tested using a Kruskal-Wallis test.

Results

The mean DM content of the haylage was 761 g/kg., CP 114 g/kg DM, NDF 568 g/kg DM and estimated DE 10.8 MJ/kg DM. The mean DM content of the hay was 929 g/kg, CP 76 g/kg DM, NDF 609 g/kg DM and estimated DE 9.8 MJ/kg DM. There was a highly significant effect of forage treatment on aerial ammonia concentration ($P<0.001$) (figure 1). There was a small, but nonsignificant effect of forage treatment on aerial ammonia concentration at 09:00 ($P>0.05$); a significant effect of forage treatment on aerial ammonia concentration at 13:00 ($P<0.05$) and a significant effect of forage treatment on aerial ammonia concentration at 17:00 ($P<0.05$) (figure 1). There were wide individual variations in ammonia concentrations within horses for both those offered haylage ($P<0.05$), and those offered hay ($P<0.01$) concentrations. According to NRC (1989) horses offered hay were estimated to be over-supplied with digestible protein by 192 g/d and those offered haylage were estimated to be over-supplied with digestible protein by 392 g/d.

Figure 1.
Effect of forage type on aerial ammonia.

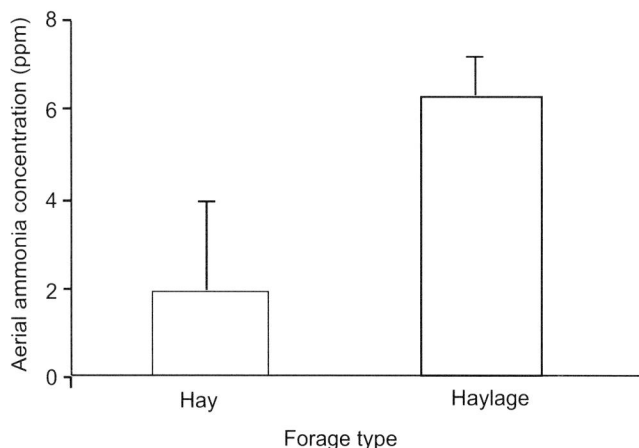

Conclusion

In conclusion, this work has shown that offering horses haylage as forage resulted in stable aerial ammonia concentrations that were significantly higher than for those offered hay as the sole forage.

Figure 2.
Effect of time
of day on
aerial
ammonia
concentrations.

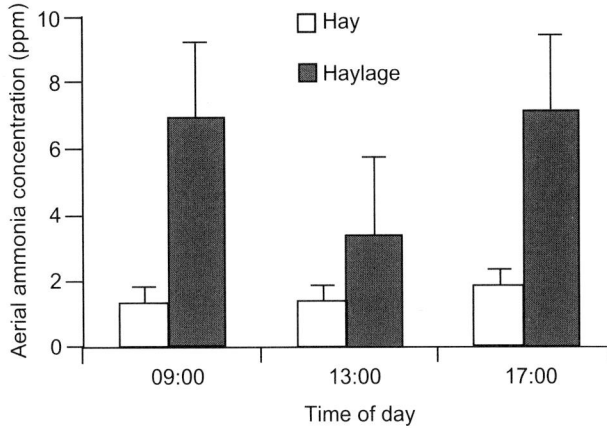

Acknowledgements

The authors are grateful to Anne and Cathy Burrell, Harrogate, North Yorkshire for the use of their stables.

References

Cuddeford, D., Woodhead, A. and Muirhead, R. (1992). A comparison between the nutritive value of short cutting cycle, high temperature dried alfalfa and timothy hay for horses. *Equine Veterinary Journal.* 24: 84-89.

Glade, M.J. (1984). The influence of dietary fiber digestibility on the nitrogen requirements of mature horses. *Journal of Animal Science.* 58: 638-646.

Moore-Colyer, M.J.S., Longland, A.C. (2000). Intakes and in vivo apparent digestibilities of four types of conserved grass forage by ponies. *Animal Science.* 71: 527-534.

NRC, (1989). *Nutritional requirements of horses.* 5th ed. Washington: National Academy Press, USA.

Pratt, S.E., Lawrence, L.M., Barnes, T., Powell, D. and Warren, L.K. (2000). Measurement of ammonia concentrations in horse stalls. *Journal of Equine Veterinary Science.* 20: 197-200.

Characterisation of the WSC fraction of six species of temperate pasture grass

J. C. Ince, [1]A. C. Longland, [1]A. J. Cairns,[1] M.Moore-Colyer [2]
[1]Institute of Grassland and Environmental Research, Plas Gogerddan, Aberystwyth, SY23 3EB, UK
[2] Institute of Rural Studies, University of Wales Aberystwyth, Llanbardan Fawr, Aberystywth, SY23 3AL, UK

Introduction

The carbohydrate (CHO) fraction of pasture grasses is a major source of energy for many domestic herbivores. However, the amounts, and types, of the water-soluble carbohydrate (WSC) fraction (i.e. glucose, fructose, sucrose, and polymers of sucrose and fructose, the fructans) present in such grasses, varies with species and environmental conditions. As the WSC constitute a highly digestible, energy yielding fraction of grasses, it is important to be able to measure their levels in a sward so that the diets of pastured animals may be designed to elicit optimal health and productivity. The aim of this study was to characterise the WSC profile of six UK pasture grasses, and to develop a technique for extracting the fructan portion of the WSC.

Materials and methods

Six species of UK pasture grasses [Cocksfoot (C), Timothy (T), Meadow Fescue (M), Italian Ryegrass (IR), Perennial Ryegrass (PR) and Hybrid Ryegrass (HR)] were grown in experimental field plots at IGER. On reaching 50% heading, grasses were harvested and immediately frozen, to prevent further CHO metabolism. Approximately 100 g of the frozen material was then oven dried to constant weight at 60 °C, and milled through a 1 mm steel mesh. Milled samples (100mg) were suspended in 1.2ml of water and placed in a boiling water bath for 5 minutes and the resultant solution analysed for total WSC content and composition (Whole grass WSC). A partially purified polymeric fructan fraction of the WSC (Fructan extract) was then prepared from the remainder of the frozen grass. The grass was washed (x 3) in boiling 80% aqueous ethanol (1:2 w/v) for 10 minutes, the liquid from each wash being retained and pooled. The grass was then boiled in water, and the washings added to the pooled aqueous ethanol washings. The total washings were reduced in volume ca. ten-fold under reduced pressure, whereupon the polymers were precipitated with 5 volumes of absolute ethanol, the monomers and dimers being retained in the liquid portion. The polymeric precipitate was harvested as a pellet *via* centrifugation at 15,000g for 15 minutes. This washing procedure was repeated (4-5 x) until the pellet was free from pigment. The Whole grass and Fructan extracts were analysed for WSC content and composition by HPLC.

Results

The whole grass contained between 17 and 30% WSC, with T and M containing the lowest and highest amounts of WSC respectively. The ryegrasses contained the highest and lowest proportions of low (monomers and dimers) and high MW WSC respectively, whereas the majority of the WSC in C, T and M was HMW fructan (Table 1). The majority of the WSC in the fructan extract of all species was HMW fructan, (mean = 68%), with relatively little glucose and fructose.

Table 1. Composition of the WSC fraction of six species of whole grass (% DM).

Sample	High MW	Raffinose	Sucrose	Glucose	Galactose	Fructose	Total WSC
C	11.036	2.098	3.007	2.041	0.296	2.610	21.278
T	10.289	1.261	1.774	1.064	-	2.005	16.775
M	15.558	2.457	4.484	2.316	-	3.998	29.735
IR	7.151	-	6.644	4.216	-	4.598	22.978
PR	10.081	-	7.709	3.733	-	4.241	26.596
HR	6.423	-	7.386	3.578	-	4.451	22.705

Table 2. Composition of the Fructan extract from six species of grass, (% DM).

Sample	High MW	Raffinose	Sucrose	Glucose	Galactose	Fructose	Total WSC
C	47.98	3.81	9.51	Tr	0.29	1	62.59
T	12.96	2.33	4.03	Tr	0.13	0.26	19.71
M	52.67	1.73	6.15	0.17	0.14	0.37	61.23
IR	38.23	10.41	17.81	3.97	0.29	2.93	73.64
PR	33.62	5.59	8.97	0.19	0.27	0.39	49.03
HR	13.09	3.49	1.69	0.86	0.27	0.4	19.8

Conclusions

All of the species tested contained substantial amounts of WSC, the composition of which varied between species. The technique for partially purifying the fructan portion of the WSC was largely successful, and the extracts of C and M contained a greater proportion of this type of material than the other species tested.

Index